HUMAN VOCAL ANATOMY

Second Printing

HUMAN VOCAL ANATOMY

———————— By ————————

DAVID ROSS DICKSON, Ph.D.
Associate Professor of Anatomy
Associate Professor of Speech
University of Pittsburgh
Pittsburgh, Pennsylvania

WILMA M. MAUE, M.A.
Assistant Director
Biocommunications Laboratory

Illustrated by
Mr. William Ronald Filer
Art Director, Medical Illustrations Department
University of Pittsburgh
School of Medicine
Pittsburgh, Pennsylvania

CHARLES C THOMAS · PUBLISHER
Springfield · Illinois · U.S.A.

Published and Distributed Throughout the World by
CHARLES C THOMAS • PUBLISHER
Bannerstone House
301-327 East Lawrence Avenue, Springfield, Illinois, U.S.A.

This book is protected by copyright. No part of it
may be reproduced in any manner without written
permission from the publisher.

© *1970, by* CHARLES C THOMAS • PUBLISHER
ISBN 0-398-00449-8
Library of Congress Catalog Card Number: 77-126473

First Printing, 1970
Second Printing, 1973

With THOMAS BOOKS careful attention is given to all details of manufacturing and design. It is the Publisher's desire to present books that are satisfactory as to their physical qualities and artistic possibilities and appropriate for their particular use. THOMAS BOOKS will be true to those laws of quality that assure a good name and good will.

Printed in the United States of America

PREFACE

The purpose of this outline is to provide the student with a concise guide to the anatomical study of those parts of the human body directly related to speech production. It is the authors' hope that this guide will serve as a complement to standard anatomical texts, atlases, and if possible, to the observation of human dissection. It should be recognized at the outset that any description of the position and attachments of muscles in the human body represents the average appearance as presently understood. We do not expect all people to be alike in external features, and yet we may be surprised to learn that internal anatomy is as variable as external features. Because of the variability of structure from one person to another, it is important to use more than one reference in the study of anatomy. Each illustration of a particular structure will differ slightly, depending on the illustrator's experience and what he is trying to highlight. All illustrations contained in this volume have been taken from dissection completed by the authors.

Work with atlases and dissection has another purpose. Learning the names and features of muscles and bones has little value unless the anatomical parts and their interrelationships are understood well enough so that they can be located, visualized, and related to the subject of our interest—the living body.

Interest in the detailed anatomy of the larynx, pharynx, palate, and tongue has been fairly recent. Knowledge of the structure and function of these areas is still incomplete. This, then, is another reason to take every opportunity to examine structures of interest in atlases, dissection, and where possible, in the living body.

Anglicized terminology has been used wherever possible. The descriptions contained in this outline do not always agree with standard descriptions contained in anatomical texts but are based on the appearance of the structures in dissections completed by the authors. Where there is question about or extreme variability in the structures or functions of anatomical parts covered, alternatives are presented. In all cases, the structures and muscle attachments presented should be taken as average findings which may not appear exactly as outlined in any given body, for there are no two bodies exactly alike.

The authors would like to express their appreciation to the many students over a period of years who, by their questions and critical comments, have contributed to the development of this work. The authors are indebted to Miss Linda Rosen, who helped edit the final copy and spent many hours organizing and completing the index.

D.R.D.
W.M.M.

CONTENTS

	Page
Preface	v
List of Illustrations	ix
I. INTRODUCTION TO THE STUDY OF HUMAN ANATOMY	3
Tissues of the Body	3
Anatomical Terms of Location and Orientation	4
II. SKELETAL FRAMEWORK	7
The Skull	7
The Vertebral Column	39
The Pelvic Girdle	42
The Shoulder Girdle	46
The Ribs	49
III. RESPIRATION	53
Muscles of the Neck	53
Muscles of the Anterior Thorax	56
Muscles of the Abdomen	64
Back Muscles	72
The Trachea and Lungs	82
Peripheral Innervation of the Muscles of Respiration	82
IV. THE LARYNX	86
The Laryngeal Cartilages	87
Ligaments of the Larynx	93
Extrinsic Muscles of the Larynx	99
Intrinsic Muscles of the Larynx	104
Peripheral Innervation of the Muscles of the Larynx	114
V. ARTICULATION	116
The Soft Palate and the Pharynx	116
Muscles of the Pharynx	117
Muscles of the Soft Palate	121
Muscles of the Mandible	128
Muscles of the Tongue	131
Muscles of the Face	139
Peripheral Innervation of the Muscles of Articulation	143
Index	145

LIST OF ILLUSTRATIONS

Figure *Page*

1. Planes of the body 5
2. Skull, lateral view showing separation of cranium and face 8
3. Skull, anterior view 9
4. Skull, lateral view 10
5. Skull, inferior view 11
6. Skull, medial view of right half of mid-sagittal section 12
7. Frontal bone, anterior view 13
8. Frontal bone, inferior view 9
9. Left parietal bone, lateral view 9
10. Occipital bone, inferior view 16
11. Occipital bone, lateral view 17
12. Left temporal bone, internal and external surfaces 12
13. Ethmoid bone, superior and lateral views 21
14. Ethmoid bone, anterior view 21
15. Sphenoid bone, anterior and superior views 23
16. Sphenoid bone, posterior view 24
17. Left maxillary bone, lateral view and nasal surface 26
18. Lacrimal bone, anterior and lateral views 28
19. Left palatine bone, posterior and medial views 29
20. Hard palate, inferior view 30
21. Vomer bone, left surface 31
22. Left zygomatic bone 32
23. Left nasal bone, external and internal surfaces 33
24. Inferior conchal bone, left medial and lateral surfaces 33

25. Mandible, external and internal surfaces ... 35
26. Hyoid bone, superior view and mid-sagittal section ... 36
27. Vertebral column, lateral view ... 37
28. Typical vertebra ... 38
29. Typical cervical vertebra ... 39
30. Atlas and Axis ... 40
31. Typical thoracic vertebra ... 41
32. Lumbar vertebra ... 42
33. Sacrum, anterior and posterior surfaces ... 43
34. Pelvic girdle, anterior view ... 44
35. Left os innominatum, lateral view ... 45
36. Shoulder girdle, left superior oblique view ... 46
37. Sternum, anterior view ... 47
38. Left clavicle, anterior view ... 48
39. Scapula, lateral and posterior views ... 49
40. Thorax, anterior view ... 50
41. Typical rib, posterior view ... 51
42. Typical articulations of ribs and vertebrae ... 52
43. Sternocleidomastoid muscle, anterior view ... 54
44. Scalene muscles, left lateral view ... 55
45. Pectoralis major muscle, anterior view ... 57
46. Pectoralis minor muscle, anterior view ... 58
47. Subclavius muscle, anterior view ... 59
48. Serratus anterior muscle, left lateral view ... 60
49. Intercostal muscles ... 62
50. Triangularis sterni muscle, posterior view ... 63
51. Abdominal wall, transverse section ... 64
52. Rectus abdominis muscle, anterior view ... 65
53. Internal oblique muscle, lateral view ... 67
54. External oblique muscle, lateral view ... 68

List of Illustrations

55. Transverse abdominal muscle, lateral view 70
56. Diaphragm, anterior view 71
57. Diaphragm, sagittal section 72
58. Trapezius muscle, posterior view 74
59. Latissimus dorsi muscle, posterior view 75
60. Serratus posterior muscles, posterior view 76
61. Ilio-costalis muscles, posterior view 78
62. Costal elevators, posterior view 80
63. Quadratus lumborum muscle, posterior lateral view 81
64. Subcostals, posterior view 83
65. Posterior view of pharynx, opened at midline 87
66. Cricoid cartilage, anterior, lateral, and posterior views 88
67. Thyroid cand cricoid cartilages 89
68. Epiglottis, posterior surface and mid-sagittal section 91
69. Arytenoid cartilage, anterolateral, dorsomedial, medial, and inferior views 92
70. Articular facet on arytenoid cartilage 92
71. Thyrohyoid membrane, lateral and anterior views 94
72. Cricothyroid membrane, superior, mid-sagittal, and anterior views 95
73. Laryngeal ligaments, posterior view 96
74. Laryngeal cartilages, mid-sagittal section 97
75. Arytenoid ligaments, mid-sagittal section and superior view 98
76. Digastric and stylohyoid muscles, lateral view with mandible cut 99
77. Suprahyoid muscles, inferior view 101
78. Suprahyoid muscles, superior posterior view 101
79. Sternohyoid and omohyoid muscles, anterior view 103
80. Extrinsic muscles of the larynx, anterior view 105
81. Cricothyroid muscle, lateral view 106
82. Posterior cricoarytenoid muscle, posterior and superior views 107
83. Posterior laryngeal muscles, posterior and superior views 108
84. Lateral cricoarytenoid muscle, lateral and superior views 110

85. Thyroarytenoid muscle, superior view ... 111
86. Transverse section of larynx ... 112
87. Frontal section of larynx ... 112
88. Larynx with thyroid lamina removed ... 114
89. Pharyngeal constrictor muscles, lateral view ... 118
90. Superior constrictor muscle, inferior view ... 119
91. Pharynx, posterior view ... 121
92. Palatopharyngeal muscles, posterior view ... 123
93. Palatopharyngeal muscles, posterior view (dissected) ... 123
94. Palatopharyngeal muscles, posterior view (dissected) ... 124
95. Extent of insertion of levator palatini muscle into the velum ... 124
96. Muscles of the velum and pharynx, lateral view ... 125
97. Palatopharyngeal muscles. (Tensor palatini muscle cut to show its attachment on Eustachian tube) ... 125
98. Mid-sagittal section of head ... 126
99. Mid-sagittal section of head (dissected) ... 127
100. Masseter muscle, lateral view ... 129
101. Pterygoid muscles, lateral and inferior views ... 130
102. Temporalis muscle, lateral views ... 132
103. Extrinsic muscles of the tongue, lateral view ... 134
104. Transverse section of the dorsum of the tongue showing fibers of the superior longitudinal muscle ... 135
105. Transverse section of the tongue ... 136
106. Transverse section of the tongue ... 137
107. Frontal section of the tongue ... 138
108. Frontal section of the tongue ... 138
109. Muscles of the face ... 140

HUMAN VOCAL ANATOMY

I

INTRODUCTION TO THE STUDY OF HUMAN ANATOMY

All living organisms are composed of **cells** which carry on the life processes of ingestion of food, metabolism, excretion, response to stimuli, motility, reproduction, and adaptation. In multicellular forms such as man, cells are differentiated into specialized forms to perform specific functions in the body. Cells which are alike in form and function are grouped together to form the **tissues** of the body. Combinations of tissues which together serve a specific body function are called **organs**. Primary life functions are carried on by groups of organs called **organ systems.**

TISSUES OF THE BODY

Bone. Dense and rigid. Forms the supporting framework of the body.

Cartilage. Like bone, but slightly or extremely flexible and softer than bone.

> **Hyaline cartilage.** Relatively rigid.
>
> **Elastic cartilage.** Nonrigid. Easily deformed.

Connective tissue. Binds together the other tissues and organs of the body.

> **Ligaments.** Slightly elastic strands of tissue which interconnect the bones and cartilages.
>
> **Membranes.** Broad, flat ligaments.
>
> **Tendons.** Inelastic strands of tissue which connect muscles to bones.
>
> **Aponeuroses.** Broad, flat tendons.
>
> **Fascia.** Bands of connective tissue which lie deep to the skin and invest the muscles and organs.

Muscle. Uniquely capable of contraction upon neural stimulation.

> **Striated muscle.** Capable of rapid, voluntary contraction. Has at-

tachments which provide for movement of body parts. An exception is heart muscle, which is involuntary.

Smooth muscle. Capable of slow, involuntary contraction. Found primarily in the digestive tract and vascular system.

Epithelial tissue. Performs the functions of protection and secretion.

Skin. Covers the external body surfaces.

Mucous membrane. Covers the internal surfaces of body cavities and passages which communicate with the exterior.

Nervous tissue. Uniquely capable of transmission of electrical impulses.

Afferent nerves. Transmit impulses toward the central nervous system.

Efferent nerves. Transmit impulses away from the central nervous system.

In the description of human anatomy, it is necessary to use language which permits accurate and consistent communication. Therefore, a few terms of anatomical position and reference must be learned initially. All descriptions will be made with reference to the body in the "anatomical position"—standing erect with arms at the sides, palms forward.

ANATOMICAL TERMS OF LOCATION AND ORIENTATION
(See Fig. 1.)

Sagittal. A vertical plane from front to back.

Coronal. A vertical plane from side to side.

Frontal. In the human, the same as **coronal**.

Transverse. A horizontal plane.

Anterior. Toward the front.

Ventral. In the human, the same as **anterior**.

Posterior. Toward the back.

Dorsal. In the human, the same as **Posterior**.

Rostral. Toward the head.

Cephalic. In the human, the same as **rostral**.

Superior. Above. In the human, the same as **rostral**.

Caudal. Toward the tail.

Inferior. Below. In the human, the same as **caudal**.

Superficial. Toward the surface.

Deep. Away from the surface.

Lateral. Toward the side.

Medial. Toward the mid-sagittal plane.

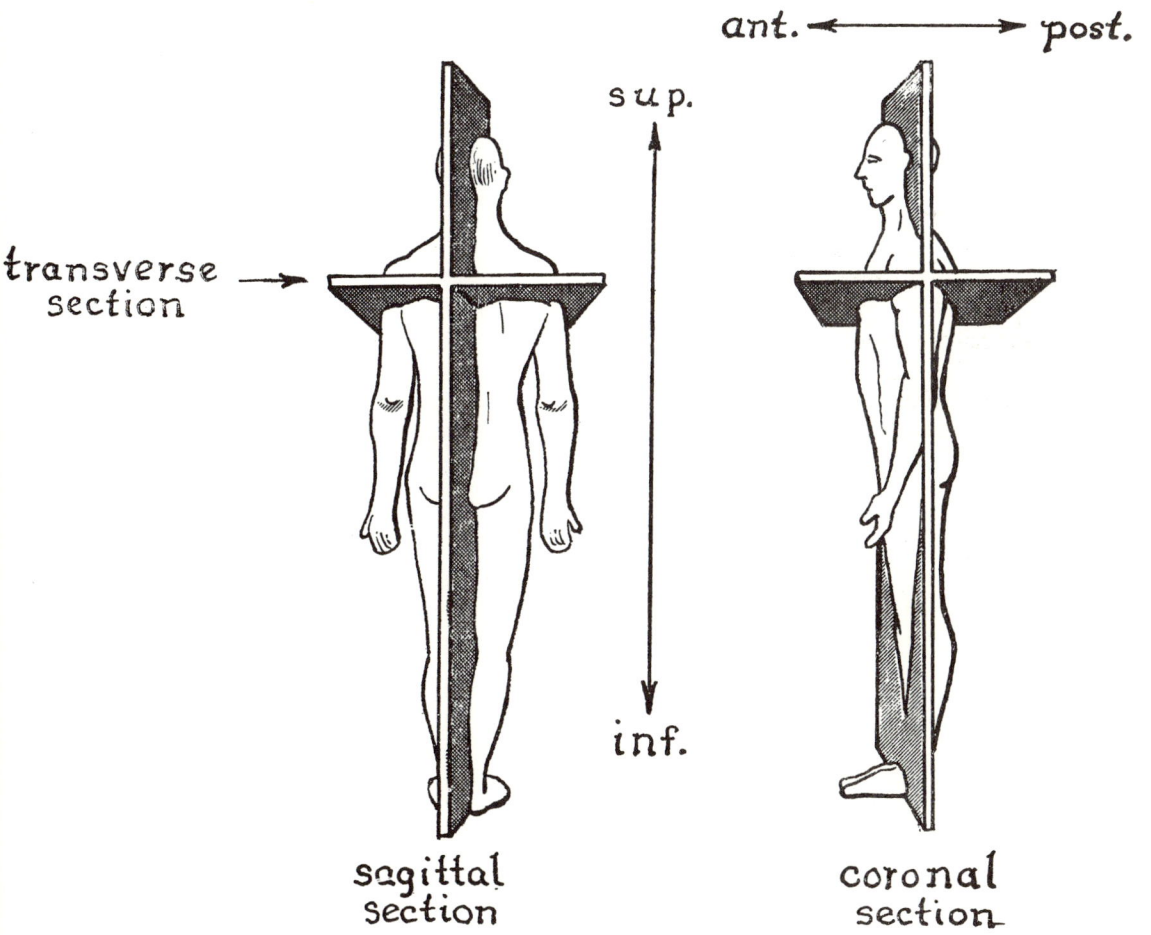

Figure 1. Planes of the body.

In the study of anatomy most terminology is Greek or Latin in origin. Since most structures are named for their appearance, location, or attachments, it is to the student's advantage to maintain a glossary of English equivalents of the Greek and Latin terms. Knowledge of definitions of terms will simplify study and make the positions and attachments of muscles clearer and easier to remember.

II
SKELETAL FRAMEWORK

The parts of the human skeleton which give rise to the muscles concerned with speech are the skull, vertebral column, shoulder girdle, pelvic girdle, and rib cage. Important landmarks to aid understanding of the form of the body and the attachments of the muscles will be indicated.

THE SKULL

The skull consists of the **cranium,** which houses the brain, and the skeleton of the **face.** The **hyoid bone,** which is located in the neck separate from the skull, will be considered with the bones of the face. (See Figs. 2-6.)

FRONTAL BONE (unpaired). Forms the anterior portion of the cranium and the vaults of the orbital cavities. (See Figs. 7 and 8.)

LANDMARKS

Ethmoid notch (unpaired). At midline between the orbital plates.

Orbital margin (paired). Anterior-superior margin of the orbital cavity.

Orbital plate (paired). Forms the vault of the orbital cavity.

Glabella (unpaired). Smooth elevation on the anterior surface between the orbital margins.

Zygomatic process (paired). At the lateral terminus of the orbital margin.

Inferior temporal line (paired). Extends superiorly and posteriorly from the zygomatic process on the external surface.

ARTICULATIONS

Inferior to the glabella with the superior margins of the **nasal bones.**

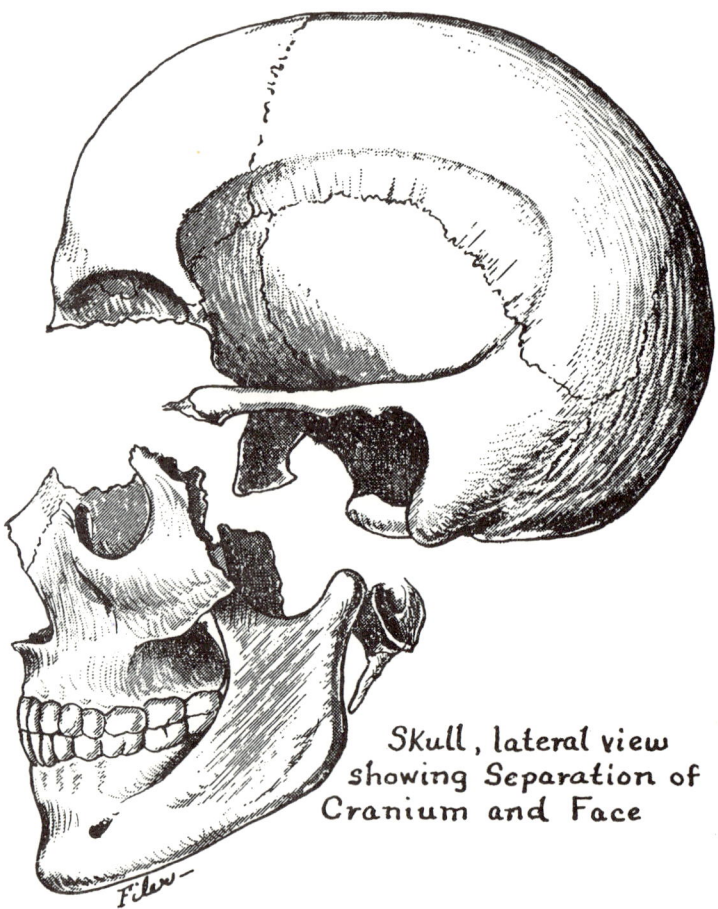

Figure 2.

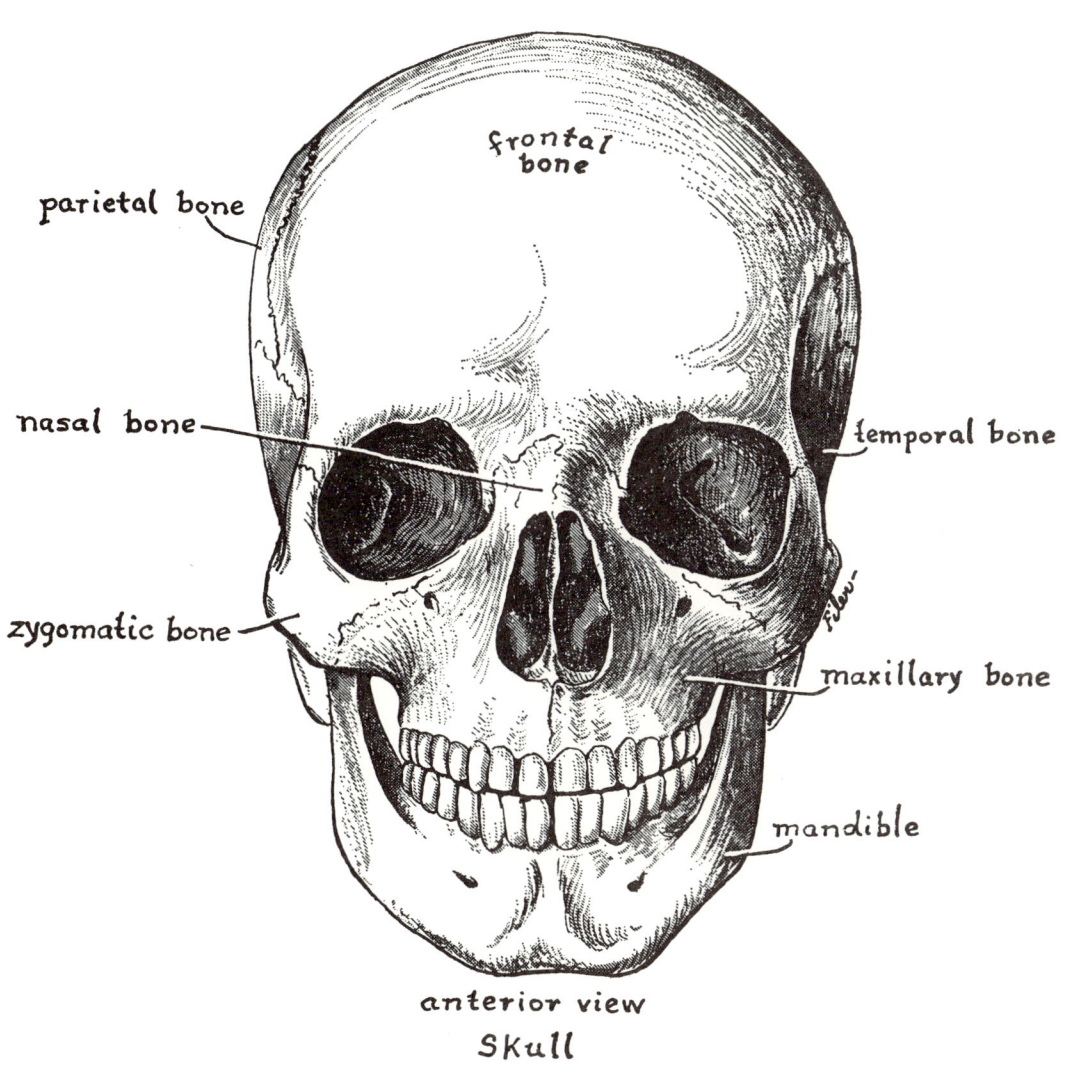

Figure 3.

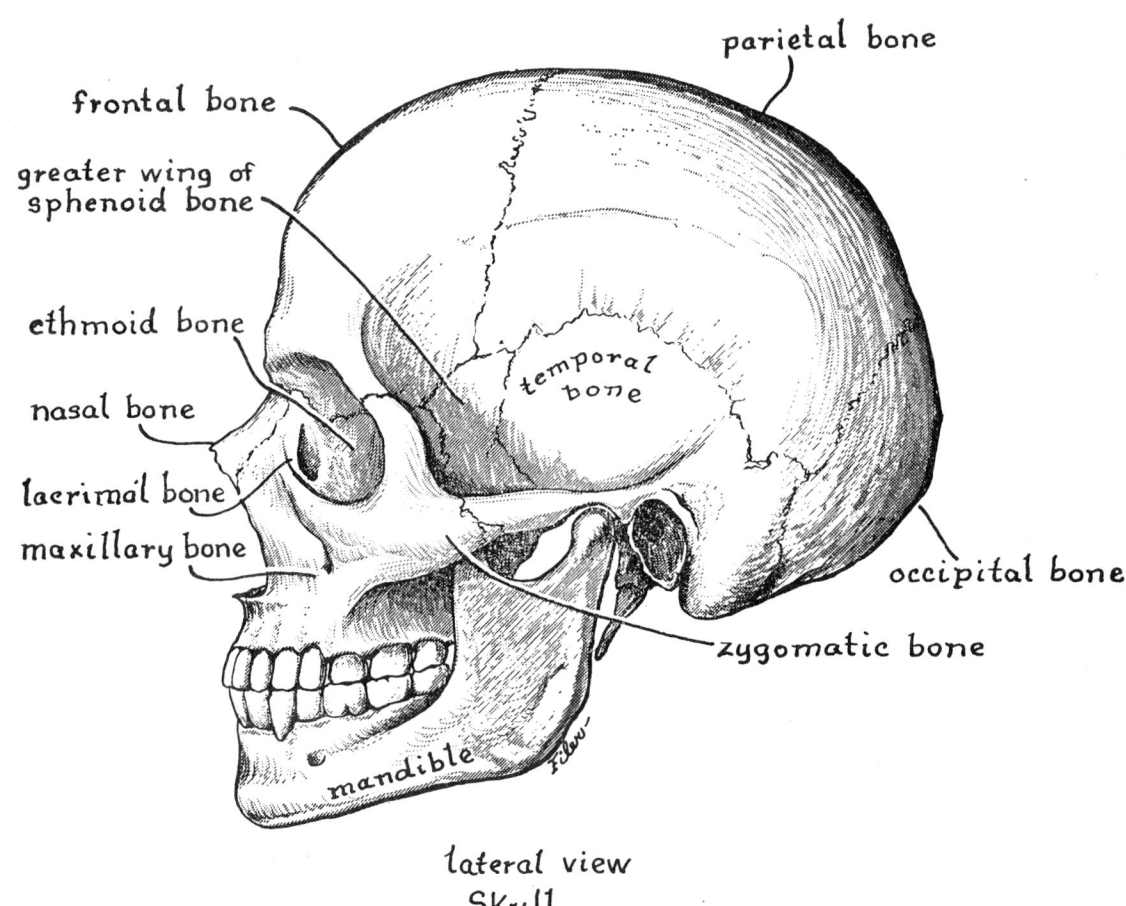

lateral view
Skull

Figure 4.

Skeletal Framework 11

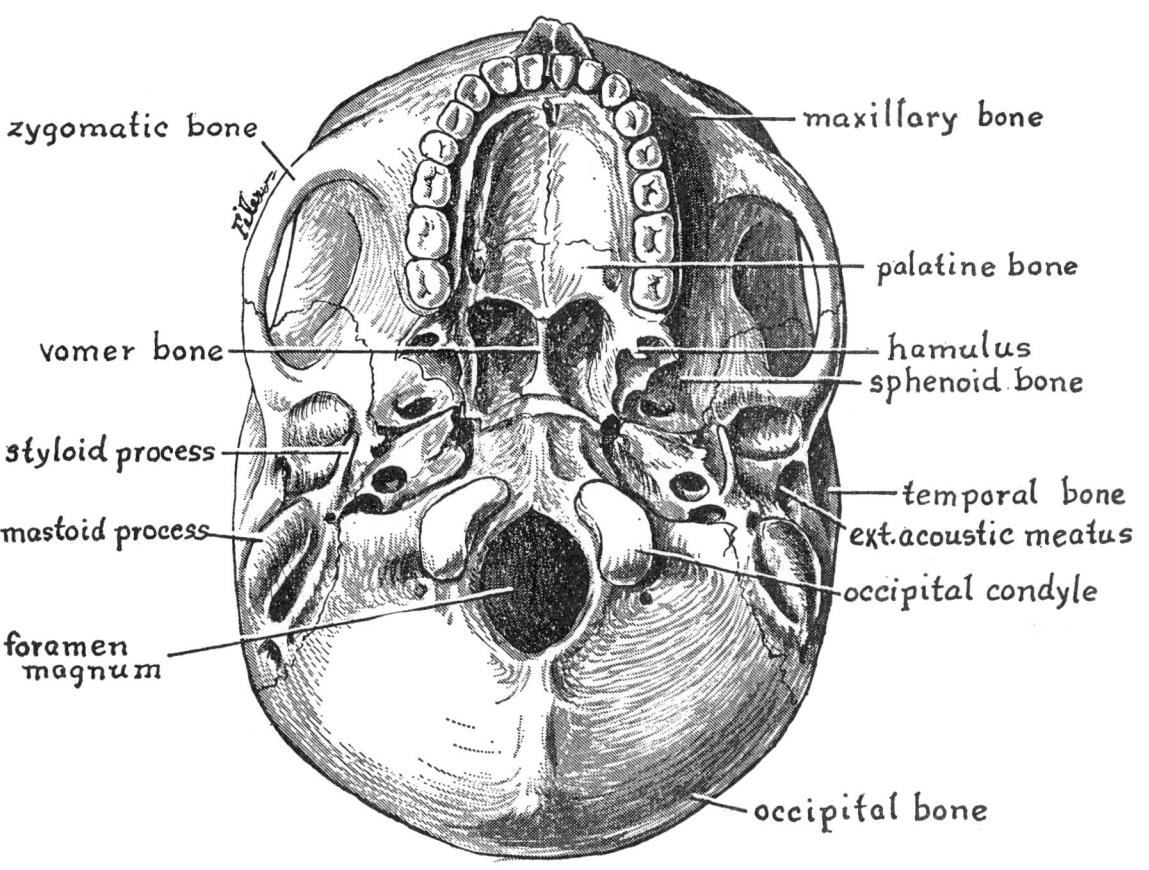

Figure 5.

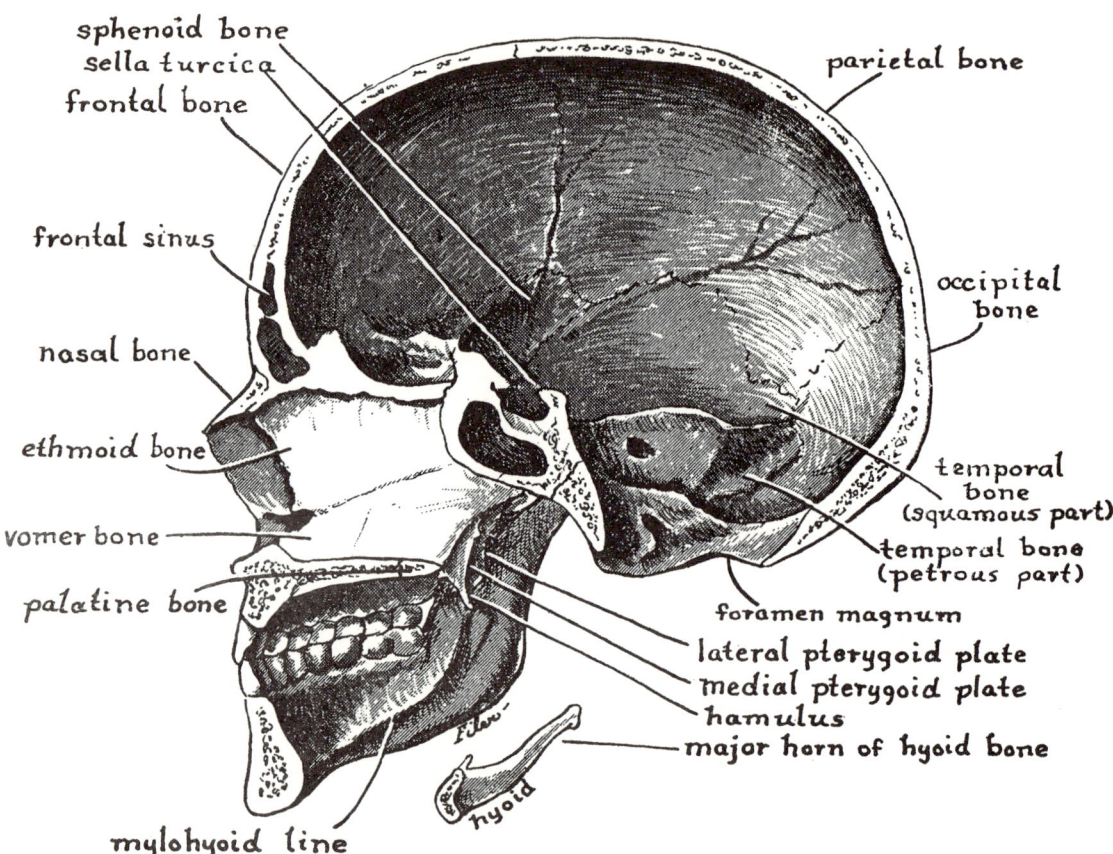

Figure 6.

Anteriorly, lateral to the nasal bones with the frontal processes of the **maxillary bones.**

Anteriorly, lateral to the maxillary bones with the superior margins of the **lacrimal bones.**

At the ethmoid notch between the orbital plates with the cribriform plate of the **ethmoid bone.**

Anterolaterally at the zygomatic processes with the frontosphenoidal processes of the **zygomatic bones.**

Laterally with the greater wings of the **sphenoid bone.**

At the posterior termini of the orbital plates with the lesser wings of the **sphenoid bone.**

Posteriorly and laterally with the anterior margins of the **parietal bones.**

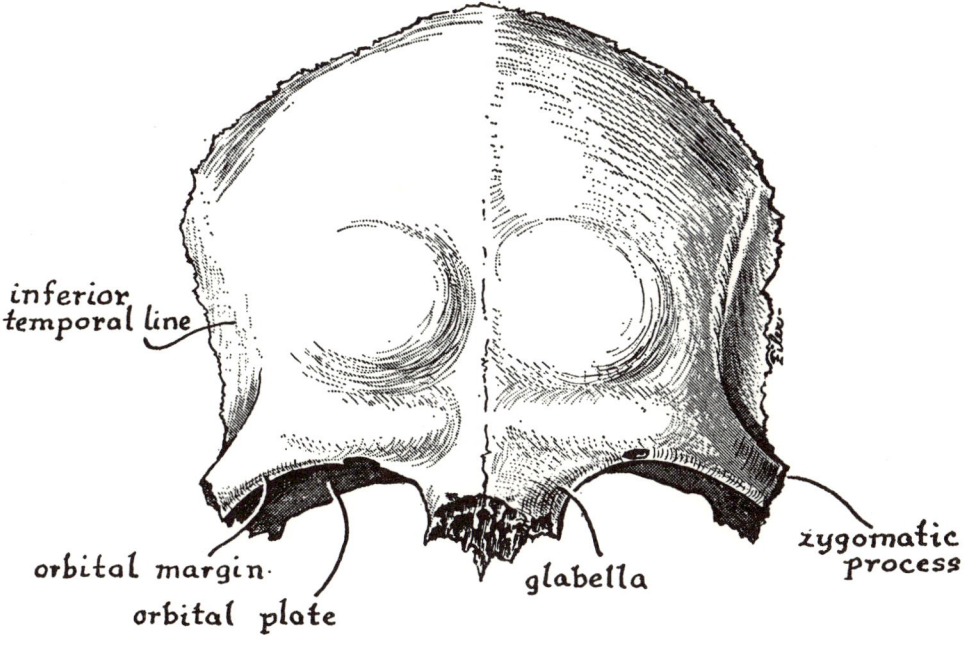

anterior view
Frontal Bone

Figure 7.

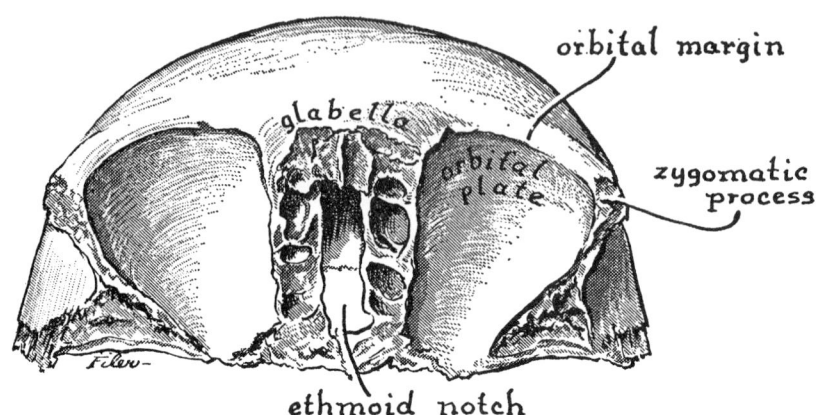

Figure 8.

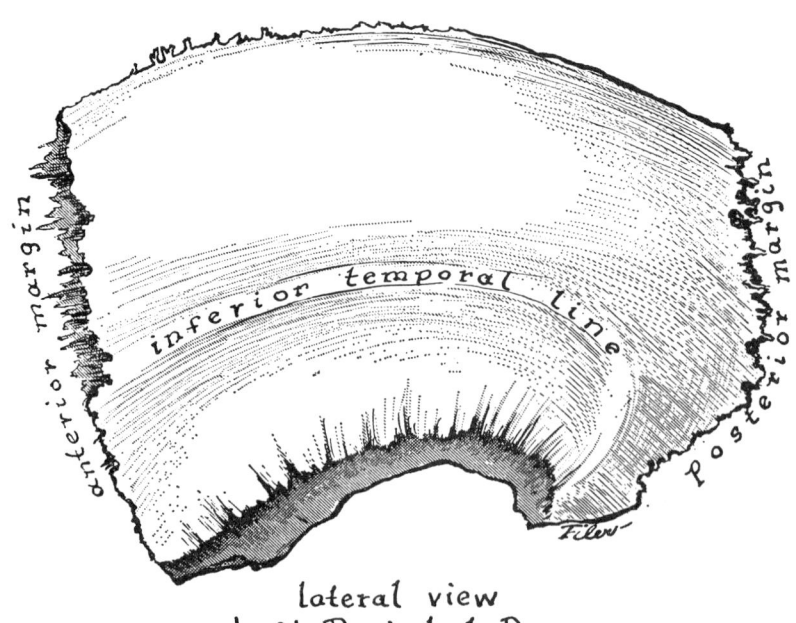

Figure 9.

PARIETAL BONE (paired). Forms the vault and part of the side wall of the cranium. (See Fig. 9.)

LANDMARKS

Inferior temporal line. From the anterior margin, curves posteriorly then inferiorly to end on the inferior border.

ARTICULATIONS

Anteriorly with the **frontal bone.**

Anterior to the temporal articulation with the greater wing of the **sphenoid bone.**

Inferiorly, at the lateral portion of the cranium with the squamous portion of the **temporal bone.**

Posteriorly with the **occipital bone.**

OCCIPITAL BONE (unpaired). Forms the posterior portion of the cranium and part of the posterior cranial floor. (See Figs. 10 and 11.)

LANDMARKS

Foramen magnum (unpaired). On the inferior surface.

Basilar portion (unpaired). Inferior part anterior to the foramen magnum.

Inferior angle (unpaired). The anterior terminus of the basilar portion.

Condyle (paired). On the inferior surface externally, lateral to the foramen magnum.

External occipital protuberance (unpaired). At midline, on the dorsal surface, midway between the superior margin and the foramen magnum.

Superior nuchal line (paired). Extends laterally from the external occipital protuberance.

ARTICULATIONS

Superiorly with the posterior borders of the **parietal bones.**

Laterally with the mastoid portions of the **temporal bones.**

Inferiorly, lateral to the basilar portion of the occipital bone with the petrosal portions of the **temporal bones.**

At the inferior angle with the body of the **sphenoid bone.**

At the condyles with the **first cervical vertebra.**

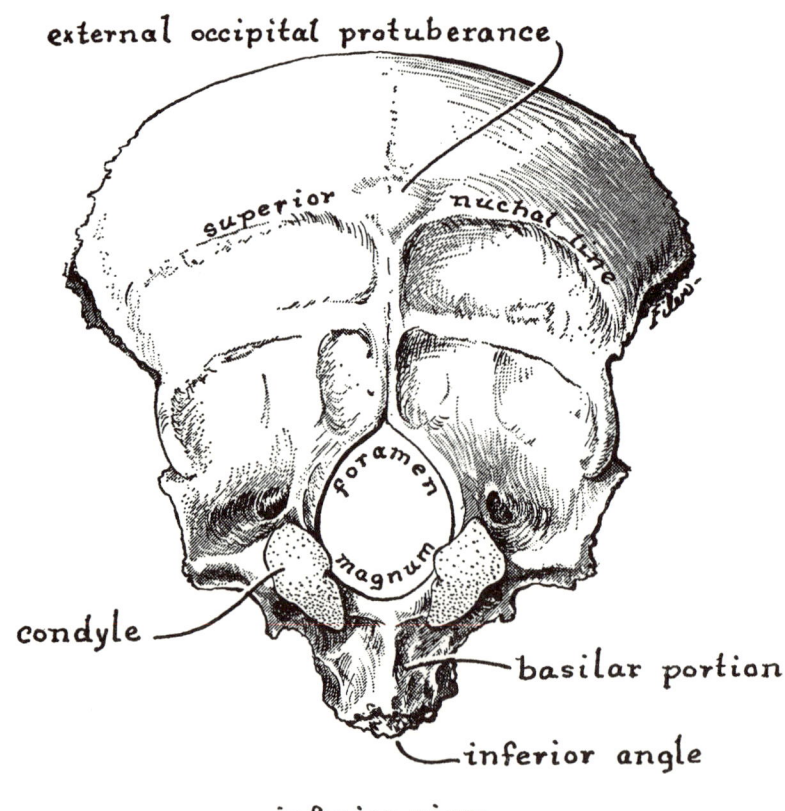

inferior view
Occipital Bone

Figure 10.

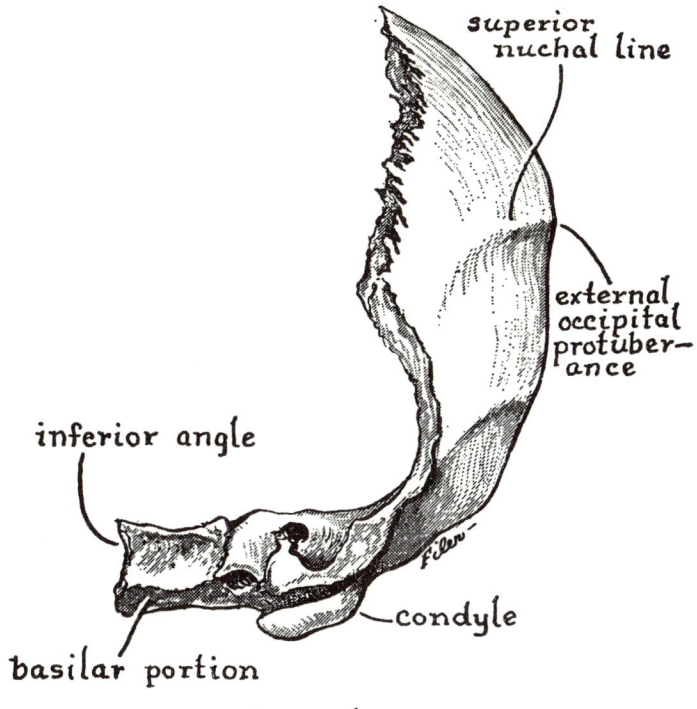

Lateral view
Occipital Bone

Figure 11.

TEMPORAL BONE (paired). At the side of the cranium, forms part of the cranial wall and base and houses the middle and inner ear. (See Fig. 12.)

LANDMARKS

    Squamous portion. Anterior-superior portion superior to the zygomatic process.

        Inferior temporal line (supramastoid line). Extends from the root of the zygomatic process horizontally across the posterior part of the squamous portion.

        Zygomatic process. Extends anteriorly from the base of the squamous portion.

        Mandibular fossa. Anterior to the external auditory meatus.

    Tympanic portion. Surrounds the external auditory meatus.

External auditory meatus. At the root of the zygomatic process.

Petrous portion. Deep to the tympanic portion. Houses the cochlea.

Mastoid portion. Posterior to the external auditory meatus.

Mastoid process. A rounded lateral projection extending antero-inferiorly from the mastoid portion.

Styloid process. Extends inferiorly, anteriorly, and slightly medially from the tympanic portion.

ARTICULATIONS

At the zygomatic process with the temporal process of the **zygomatic bone.**

Anteriorly with the greater wing of the **sphenoid bone.**

Superiorly with the inferior border of the **parietal bone.**

Medially on the inferior surface with the **occipital bone.**

At the root of the zygomatic process with the condyloid process of the **mandible.**

ETHMOID BONE (unpaired). In midline, forms part of the cranial base and part of the skeleton surrounding the nasal cavities. (See Figs. 13 and 14.)

LANDMARKS

Cribriform plate (unpaired). Horizontal plate forming the vaults of the nasal cavities and part of the cranial floor.

Perpendicular plate (unpaired). Vertical plate in midline forming part of the nasal septum.

Lateral mass (paired). The bony labyrinth lateral to the nasal cavity.

Lamina papyracea (paired). Lateral wall of the lateral mass forming part of the medial wall of the orbital cavity.

Skeletal Framework

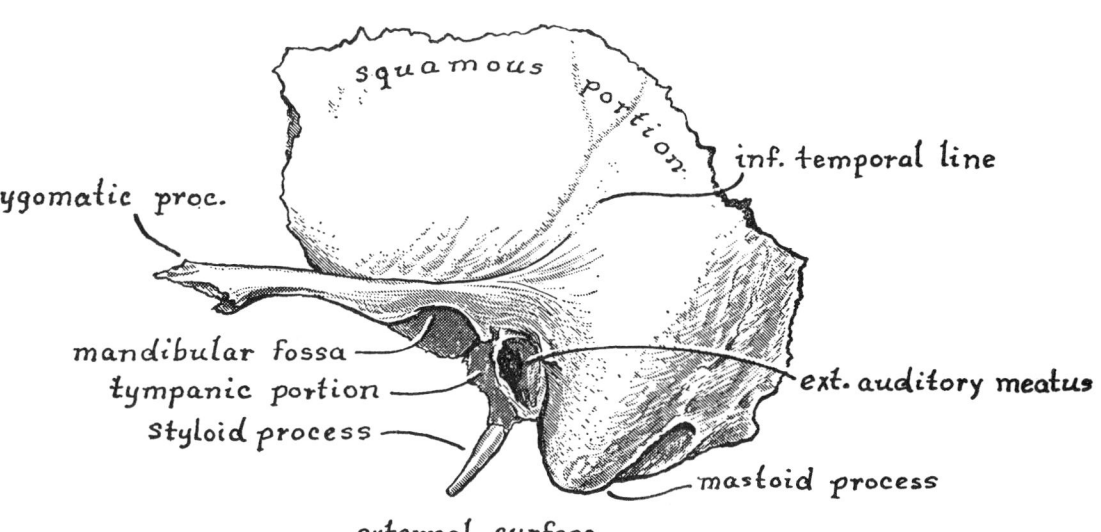

external surface

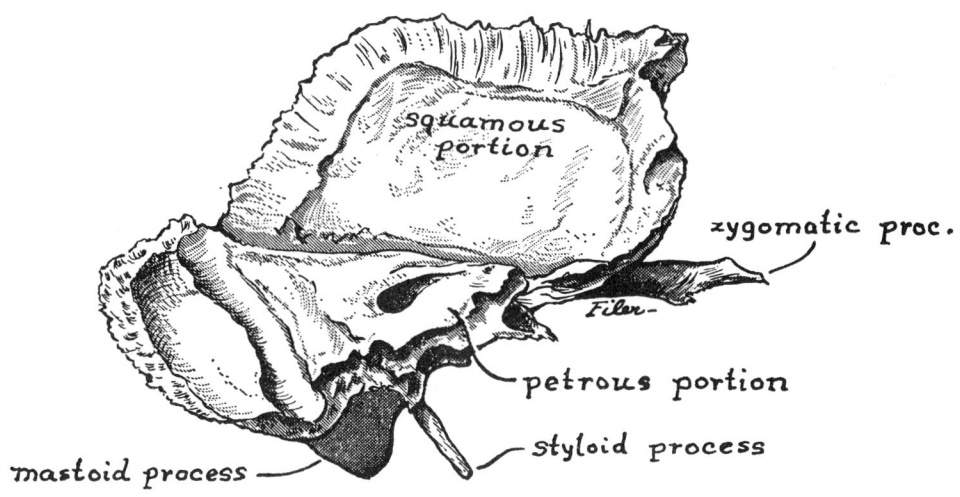

internal surface
Left Temporal Bone

Figure 12.

Uncinate process (paired). Projects inferiorly from beneath the superior concha for articulation with the inferior conchal bone.

Superior concha (paired). Convoluted projection into the nasal cavity from the medial wall of the lateral mass.

Medial concha (paired). Projects medially from the lateral mass just below the superior concha.

ARTICULATIONS

Anterior and lateral to the cribriform plate with the ethmoid notch of the **frontal bone.**

In midline at the anterior tip of the perpendicular plate with the deep surfaces of the **nasal bones.**

At the anterior borders of the laminae papyraceae with the posterior borders of the **lacrimal bones.**

At the uncinate processes with the **inferior conchal bones.**

At the inferior borders of the laminae papyraceae with the orbital surfaces of the **maxillary bones.**

At the posterior surfaces of the lateral masses with the ethmoid crests and surfaces of the **palatine bones.**

At the posterior inferior border of the perpendicular plate with the **vomer bone.**

SPHENOID BONE (unpaired). Forms part of the base and lateral walls of the cranium and the vault of the pharynx. (See Figs. 15 and 16.)

LANDMARKS

Body (unpaired). Central mass at midline of the cranial base.

Sella turcica (unpaired). Depression in the superior surface of the body.

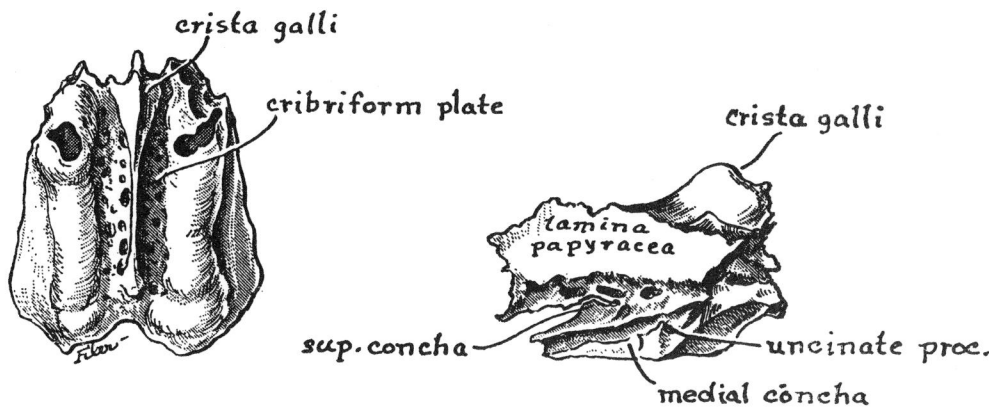

Figure 13.

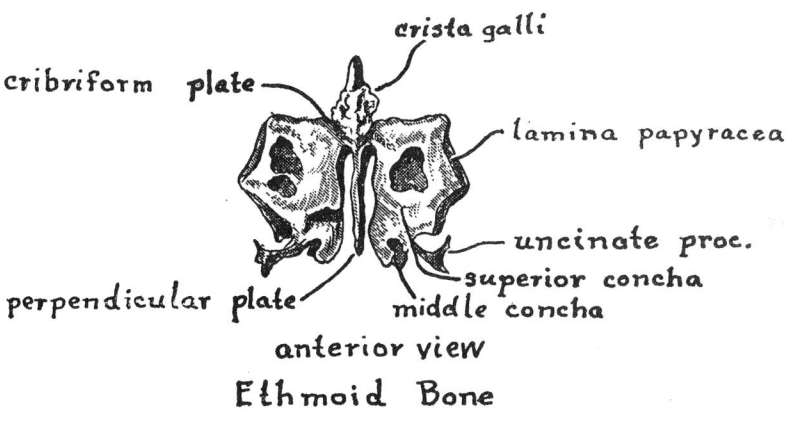

Figure 14.

Greater wing (paired). Extends laterally from the body, then curves superiorly to form part of the lateral cranial wall.

Orbital surface (paired). Medial surface of the superior extension of the greater wing.

Infratemporal crest (paired). The line of flexion of the greater wing as it bends superiorly at the side of the cranium.

Petrosal process (paired). Posterior pointed extension of the greater wing on the cranial base between the squamosal and petrosal portions of the temporal bone.

Angular spine (paired). Extends inferiorly at the most posterior point of the petrosal process.

Lesser wing (paired). Extends laterally from the superior margin of the body. Forms part of the cranial base.

Pterygoid process (paired). Extends inferiorly from the body at the root of the greater wing, then divides to form the medial and lateral pterygoid plates.

Medial pterygoid plate (paired). Medial extension of the pterygoid process. Forms the posterior portion of the lateral wall of the nasal cavity.

Lateral pterygoid plate (paired). Lateral extension of the pterygoid process.

Hamulus (paired). Thin, hooklike inferior projection from the medial pterygoid plate slightly inferior and posterior to the hard palate.

Pterygoid fossa (paired). Depression posteriorly between the medial and lateral pterygoid plates.

Scaphoid fossa (paired). Depression in the posterior surface of the pterygoid process superior to the pterygoid fossa.

ARTICULATIONS

At the anterior borders of the greater and lesser wings with the orbital surfaces of the **frontal bone.**

At the superior borders of the greater wings with the inferior borders of the **parietal bones.**

At the posterior borders of the greater wings and the medial borders of

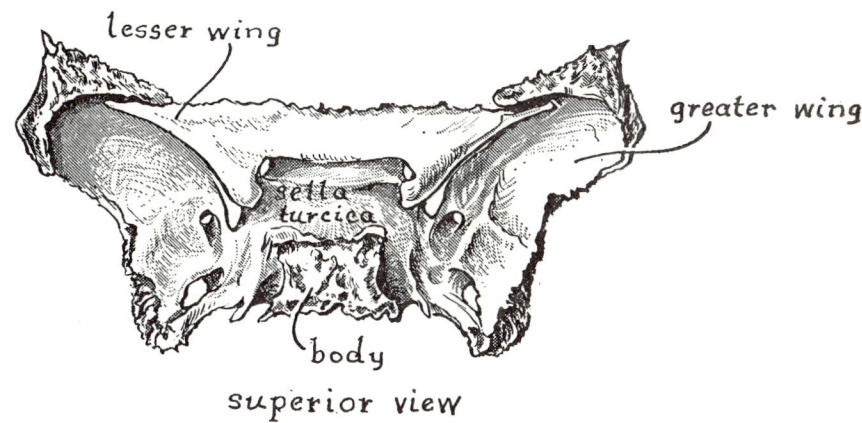
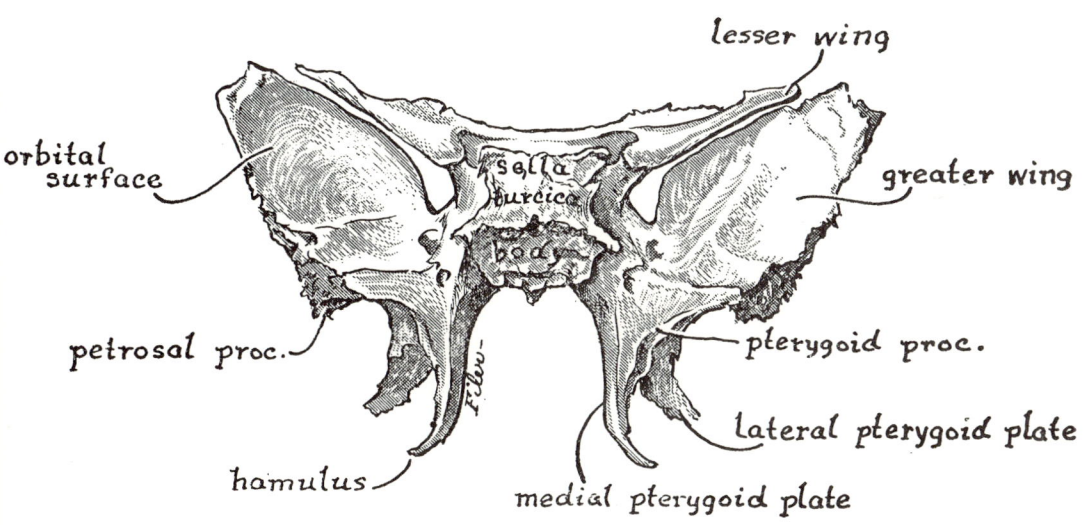

Figure 15.

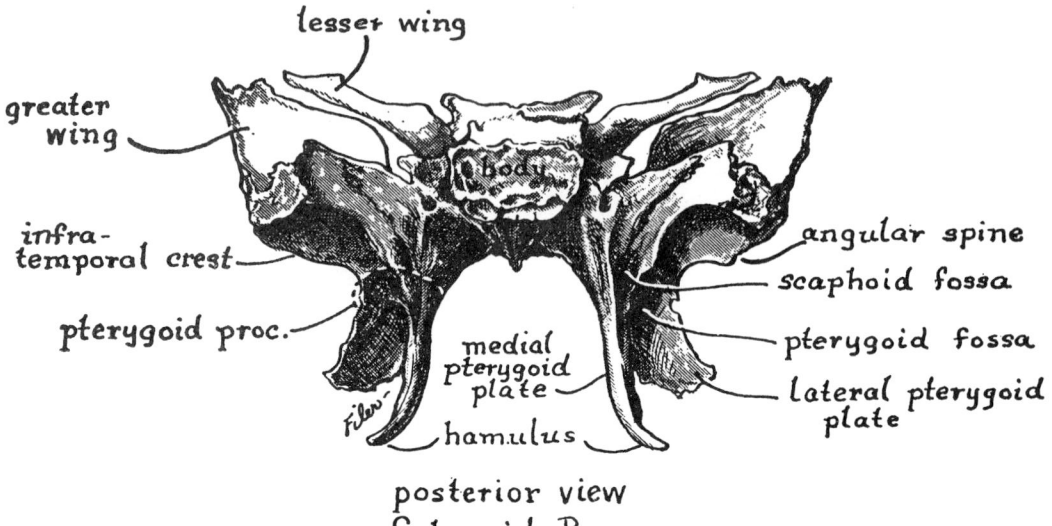

Figure 16.

the petrosal processes with the squamous and petrosal portions of the **temporal bones.**

At the posterior surface of the body with the inferior angle of the **occipital bone.**

At the anterior borders of the lateral ptergygoid plates with the tuberosities of the **maxillary bones** (inconstant).

At the lateral margins of the orbital surfaces of the greater wings with the frontosphenoidal processes of the **zygomatic bones.**

Anteriorly at midline with the cribriform and perpendicular plates of the **ethmoid bone.**

At midline on the inferior surface of the body with the posterior superior part of the **vomer bone.**

At the anterior borders of the medial and lateral pterygoid plates with the pyramidal and sphenoidal processes of the **palatine bones.**

MAXILLARY BONE (paired). Forms the upper jaw. (See Fig. 17.)

LANDMARKS

Body. Roughly pyramidal anterior mass of the maxilla.

>Facial surface. Anterior surface.

>>Frontal process. Extends superiorly from the facial surface near midline.

>>Nasal notch. Extends laterally from the midline below the frontal process.

>>Anterior nasal spine. Projects anteriorly at midline below the nasal notch.

>>Zygomatic process. Extends laterally from the facial surface to form part of the zygomatic arch.

>>Canine fossa. Superior to the lateral canine tooth on the facial surface.

>Orbital surface. Superior surface. Forms part of the orbital floor.

>Nasal surface. Medial surface. Forms part of the lateral wall of the nasal cavity.

>>Ethmoid crest. On the superior part of the nasal surface of the frontal process.

>>Conchal crest. On the inferior part of the nasal surface of the frontal process.

>Infratemporal surface. Posterior surface.

>>Maxillary tuberosity. At the inferior boundry of the infratemporal surface.

>Palatine process. With the process of the other side, forms the anterior three fourths of the hard palate.

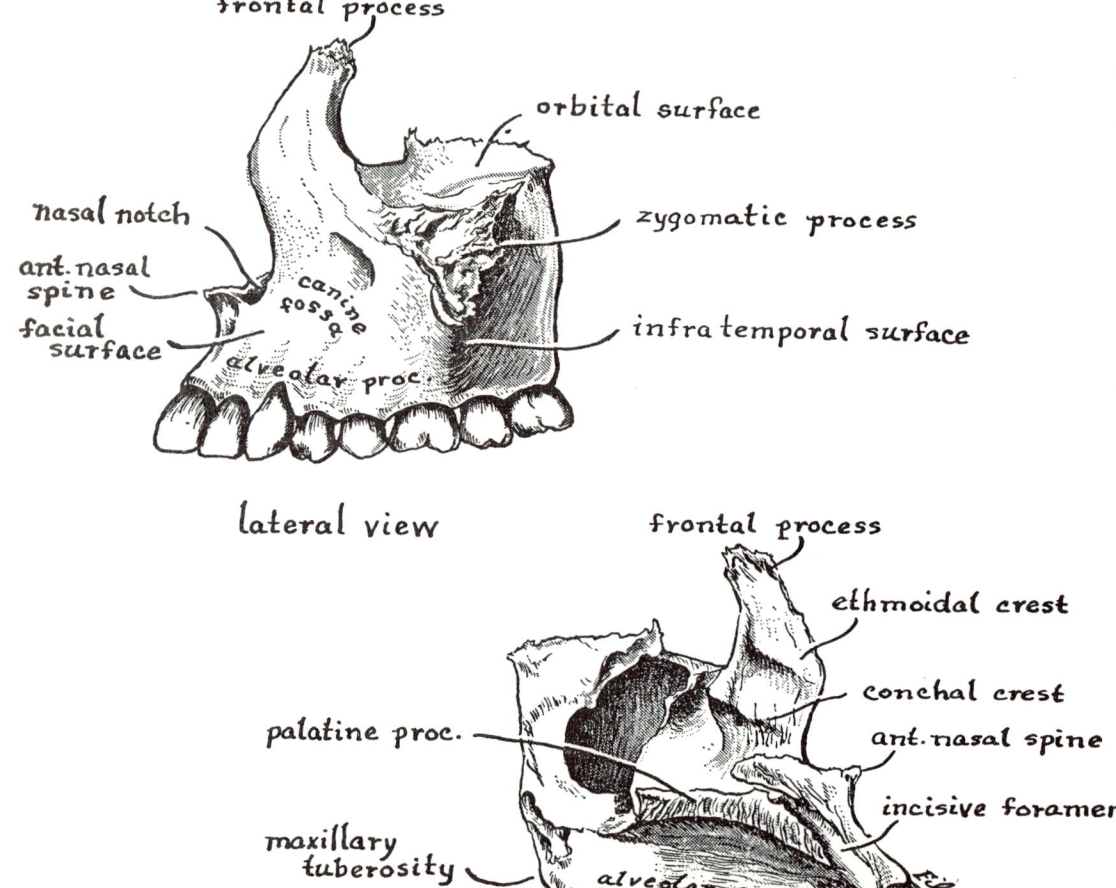

Left Maxillary Bone

Figure 17.

Alveolar process. Extends inferiorly from the body to house the roots of the maxillary teeth.

Incisive foramen. In midline on the hard palate posterior to the central incisor teeth.

Premaxilla. The small anterior portion of the hard palate bounded by lines from the incisive foramen to points between the lateral incisors and the canine teeth.

ARTICULATIONS

At the superior border of the frontal process with the **frontal bone.**

At the medial border of the frontal process with the lateral border of the **nasal bone.**

At the medial border of the orbital surface with the lamina papyracea of the **ethmoid bone.**

At the medial border of the orbital surface with the anterior and inferior portions of the **lacrimal bone.**

Laterally at the zygomatic process and at the orbital surface with the maxillary and orbital processes of the **zygomatic bone.**

At the superior surface of the palatine process at midline with the inferior border of the **vomer bone.**

At the posterior border of the palatine process and the posterior portion of the nasal surface with the anterior parts of the **palatine bone.**

At the conchal crest with the **inferior conchal bone.**

LACRIMAL BONE (paired). Located on the medial surface of the orbital cavity between the frontal process of the maxilla and the lamina papyracea of the ethmoid bone. (See Fig. 18.)

LANDMARKS

Orbital surface. Lateral surface.

Nasal surface. Medial surface.

ARTICULATIONS

Superiorly with the **frontal bone.**

Anteriorly and inferiorly with the frontal process of the **maxillary bone.**

At the anterior-inferior extremity with the **inferior conchal bone.**

Posteriorly with the lamina papyracea of the **ethmoid bone.**

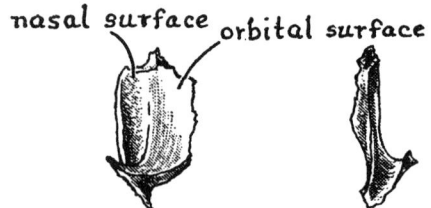

Figure 18.

PALATINE BONE (paired). An "L"-shaped bone which forms the posterior one fourth of the hard palate and part of the lateral wall of the nasal cavity posteriorly. (See Figs. 19 and 20.)

LANDMARKS

Horizontal part. Forms the posterior one fourth of the hard palate with the part from the other side.

  Posterior nasal spine. Projects posteriorly at midline at the junction of the horizontal parts from the two sides.

Vertical part. Forms part of the lateral wall of the nasal cavity posteriorly.

  Orbital process. Most superior extension of the vertical part.

Ethmoidal surface. Nasal surface of the orbital process.

Ethmoidal crest. Superiorly on the nasal surface of the vertical part.

Sphenoid process. At the posterior end of the ethmoidal crest.

Conchal crest. Midway on the nasal surface of the vertical part.

Pyramidal process. Inferior half of the posterior border of the vertical part.

ARTICULATIONS

Along the anterior border and the anterior part of the lateral surface with the palatine process and nasal surface of the **maxillary bone.**

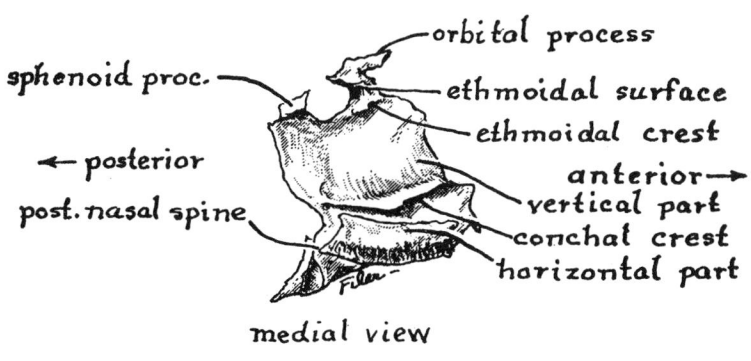

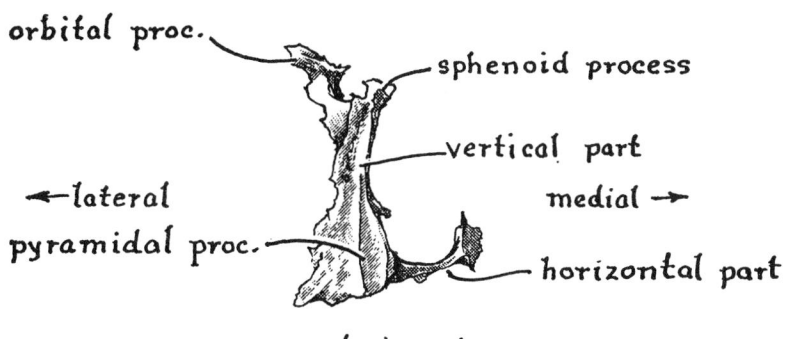

Figure 19.

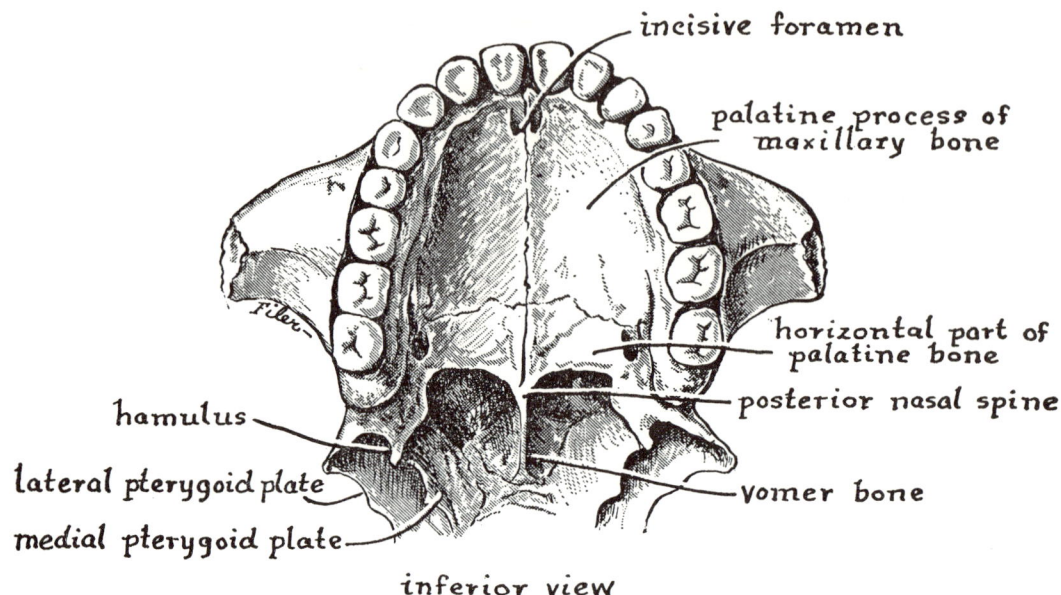

Figure 20.

At the ethmoid crest and ethmoidal surface with the posterior surface of the lateral mass of the **ethmoid bone**.

At the conchal crest with the **inferior conchal bone**.

At the sphenoid process and the superior border of the horizontal part at midline with the posterior fourth of the inferior border of the **vomer bone**.

At the pyramidal process with the pterygoid process of the **sphenoid bone**.

VOMER BONE (unpaired). Flat, plowshare shaped bone. Forms the posterior and inferior part of the nasal septum. (See Fig. 21.)

ARTICULATIONS

At the upper half of the anterior border with the perpendicular plate of the **ethmoid bone**.

At the superior border with the body of the **sphenoid bone.**

At the anterior three fourths of the inferior border with the palatine processes of the **maxillary bones.**

At the posterior one fourth of the inferior border with the horizontal part of the **palatine bones.**

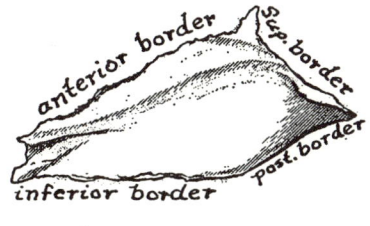

left surface
Vomer Bone

Figure 21.

ZYGOMATIC BONE (paired). Forms the bony prominence of the cheek and part of the lateral wall and floor of the orbital cavity. (See Fig. 22.)

LANDMARKS

Frontosphenoidal process. Superior projection.

Maxillary process. Inferior medial projection.

Temporal process. Lateral posterior projection. Forms the anterior part of the zygomatic arch.

Orbital process. Posterior projection between the frontosphenoidal process and the maxillary process. Forms part of the floor and lateral wall of the orbital cavity.

ARTICULATIONS

At the frontosphenoidal process with the zygomatic process of the **frontal bone.**

At the orbital process with the greater wing of the **sphenoid bone.**

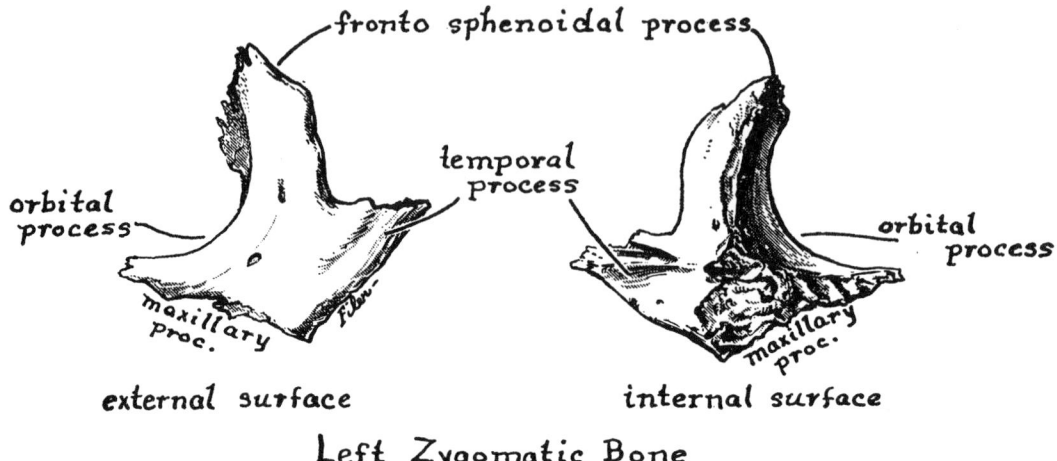

Left Zygomatic Bone

Figure 22.

At the *maxillary and orbital processes* with the zygomatic process and orbital surface of the **maxillary bone.**

At the *temporal process* with the zygomatic process of the **temporal bone.**

NASAL BONE (paired). At midline anteriorly. Forms the bridge of the nose between the orbital cavities. (See Fig. 23.)

LANDMARKS

Vertical crest. Posterior extension of the medial border.

ARTICULATIONS

At the *superior margin* with the **frontal bone.**

At the *lateral border* with the frontal process of the **maxillary bone.**

At the *vertical crest* with the perpendicular plate of the **ethmoid bone.**

INFERIOR CONCHAL BONE (paired). Also called the inferior turbinated bone. Extends horizontally along the lateral wall of the nasal cavity inferior to the medial concha of the ethmoid bone. (See Fig. 24.)

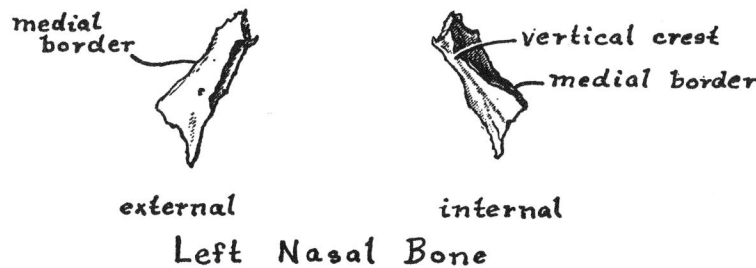

Left Nasal Bone

Figure 23.

LANDMARKS

Lacrimal process. Extends anteriorly from the superior border.

Maxillary process. Extends inferiorly from the posterior superior surface.

Ethmoid process. Superior to the maxillary process.

ARTICULATIONS

Posteriorly and superiorly with the conchal crest of the **palatine bone.**

At the lacrimal process with the anterior inferior extension of the **lacrimal bone.**

At the maxillary process with the conchal crest of the **maxillary bone.**

At the ethmoid process with the uncinate process of the **ethmoid bone.**

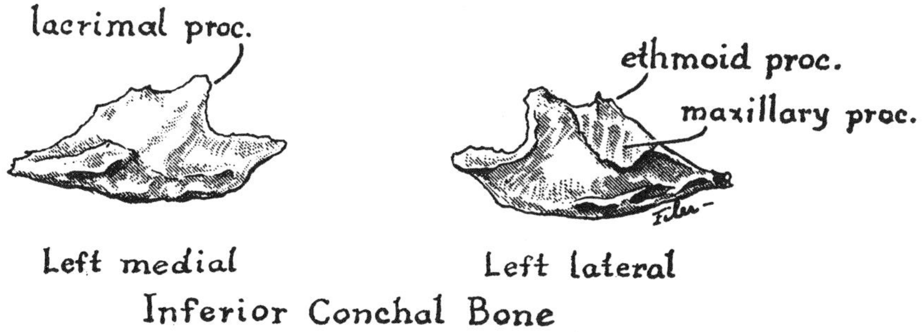

Inferior Conchal Bone

Figure 24.

MANDIBLE (unpaired). Forms the lower jaw. (See Fig. 25.)

LANDMARKS

    Body (unpaired). Long curved horizontal portion in the shape of a horseshoe.

        Symphysis (unpaired). The union of the left and right halves of the body anteriorly at midline.

        Oblique line (paired). Ridge on the external surface of the body extending from below the second premolar tooth, posteriorly and superiorly, to become continuous with the anterior border of the ramus.

        Incisive fossa (paired). On the external surface of the body below the incisor teeth lateral to the symphysis.

        Superior mental spine (paired). Variable. A small prominence on the deep surface of the body lateral to the symphysis.

        Inferior mental spine (paired). Variable. Just below the superior mental spine.

        Alveolar process (paired). Projects superiorly from the body to house the roots of the mandibular teeth.

        Mylohyoid line (paired). Roughly horizontal ridge on the deep surface of the body. Extends beneath the molar teeth from near the symphysis.

        Angle (paired). The most posterior-inferior point of the mandible at the junction of the body and ramus.

    Ramus (paired). A large process which projects superiorly from the posterior end of the body.

        Condyle (paired). The most superior-posterior projection from the ramus. Articulates with the temporal bone to form the temporomandibular joint.

Coronoid process (paired). Superior projection from the ramus anterior to the condyle.

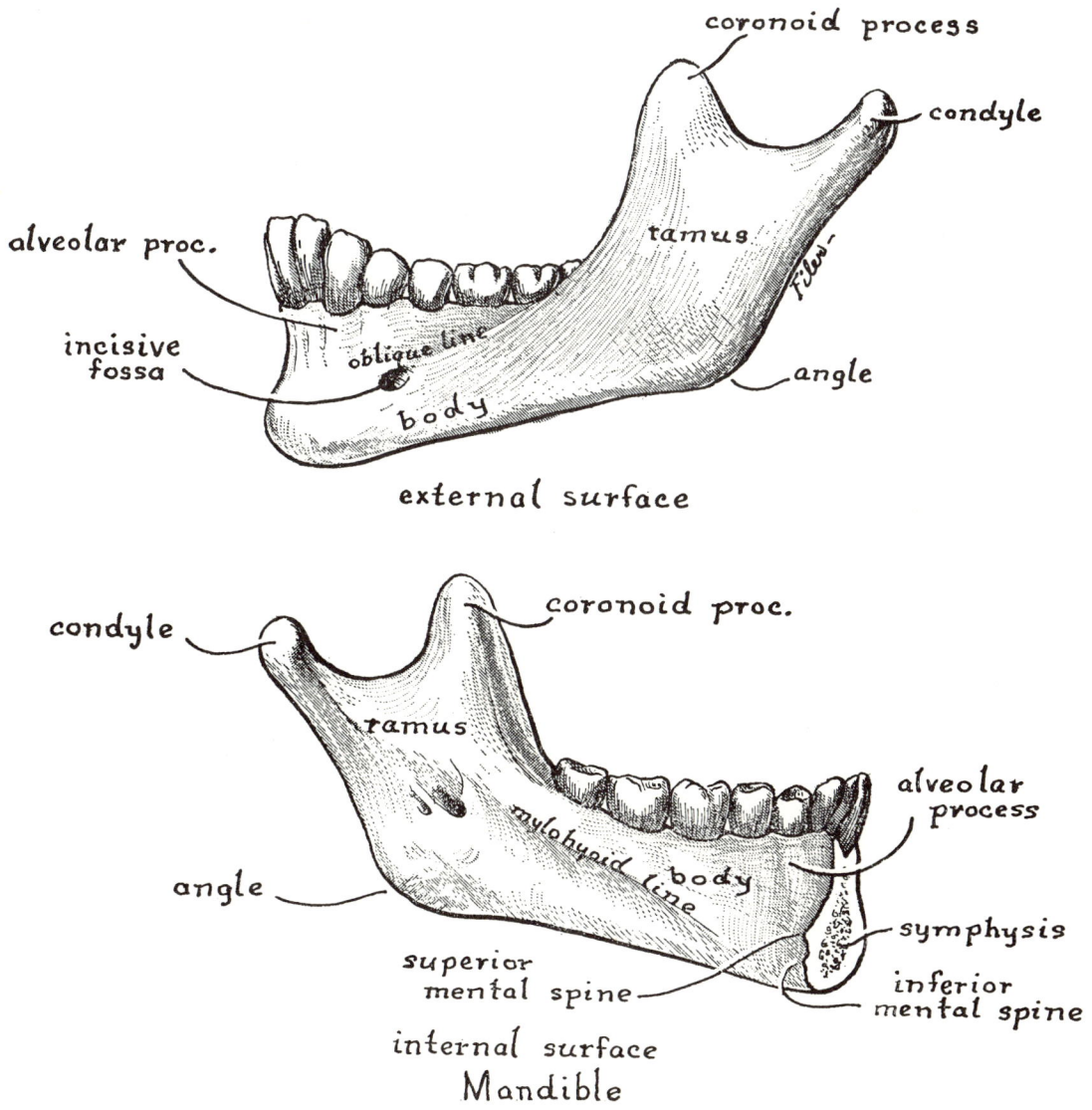

Figure 25.

HYOID BONE (unpaired). Horseshoe-shaped bone in the neck above the larynx. Has no bony articulations. (See Fig. 26.)

LANDMARKS

Body (unpaired). Rounded anterior portion.

Major cornu (paired). Posterior extension from the side of the body.

Minor cornu (paired). Short superior projection from the junction of the body and major cornu.

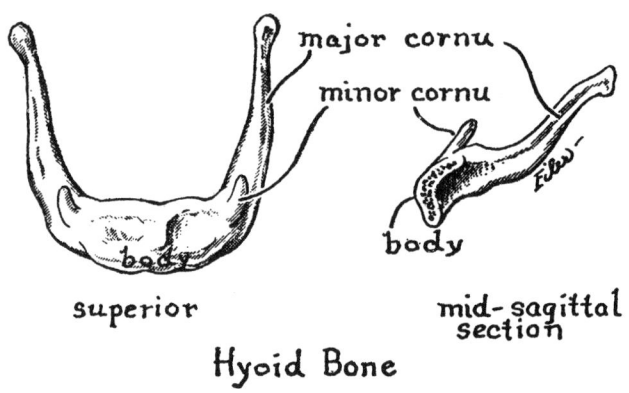

Figure 26.

THE VERTEBRAL COLUMN

The vertebral column is the flexible backbone of the trunk and neck and is composed of five major divisions which are named for their location. The seven unpaired **cervical vertebrae** are located in the neck. The twelve unpaired **thoracic vertebrae** are located at the back of the rib cage. The five unpaired **lumbar vertebrae** are located between the rib cage and the pelvis. The five fused unpaired **sacral vertebrae** form the back of the pelvic girdle. From three to five small unpaired bones inferior to the sacrum form the **coccyx**, or rudimentary tail. (See Fig. 27.)

TYPICAL VERTEBRAE. (See Fig. 28.)

Body (unpaired). Rounded anterior portion.

Neural arch (unpaired). Surrounds the vertebral foramen posterior to the body.

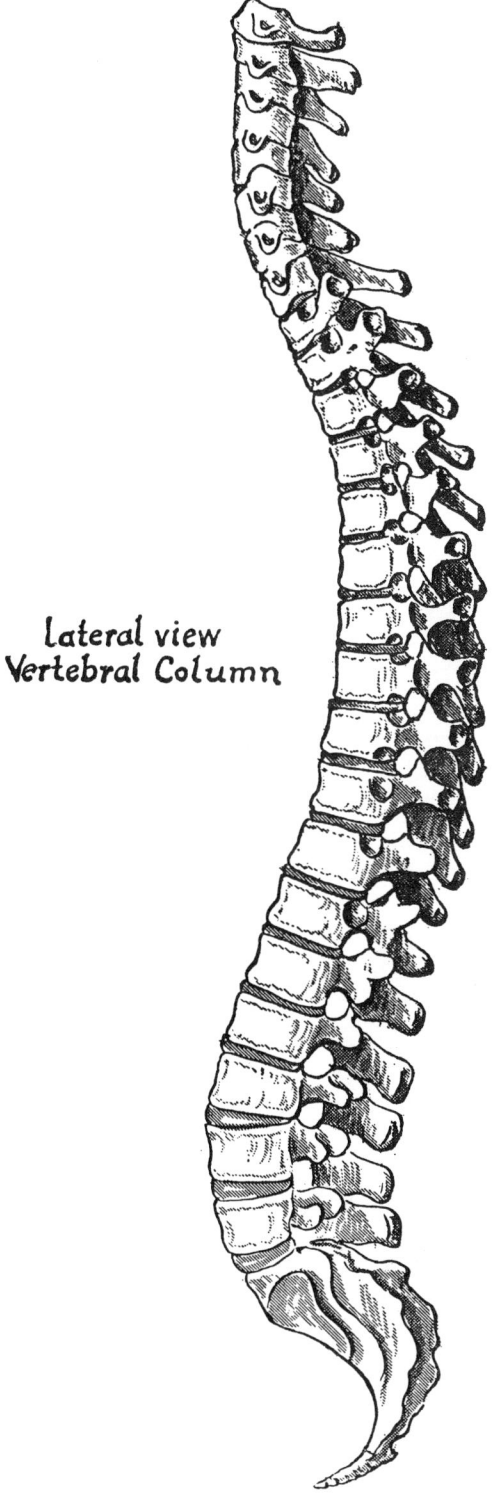

Figure 27.

Pedicle (paired). Anterolateral part of the neural arch. Extends posteriorly from the dorsolateral part of the body.

Lamina (paired). Posterolateral part of the neural arch. Extends posteriorly from the pedicle. The two laminae join posteriorly.

Vertebral foramen (unpaired). Central foramen through which the spinal cord passes. Surrounded by the neural arch.

Transverse process (paired). Extends laterally from the junction of the pedicle and lamina.

Spinous process (unpaired). Extends posteriorly and inferiorly from the junction of the two laminae.

Superior articular process (paired). Extends superiorly from the junction of the pedicle and lamina. Has an articular facet which faces posteriorly to articulate with the vertebra above.

Inferior articular process (paired). Extends inferiorly from the lamina. Has an articular facet which faces anteriorly to articulate with the vertebra below.

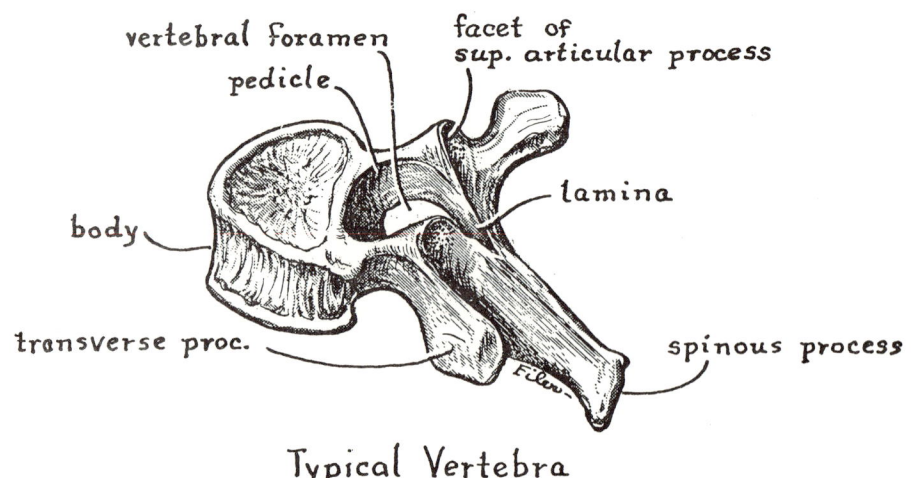

Figure 28.

TYPICAL CERVICAL VERTEBRAE. (See Fig. 29.)

Body (unpaired). Tends to be small.

Transverse process (paired). Perforated by transverse foramen.

Spinous process (unpaired). May be bifid.

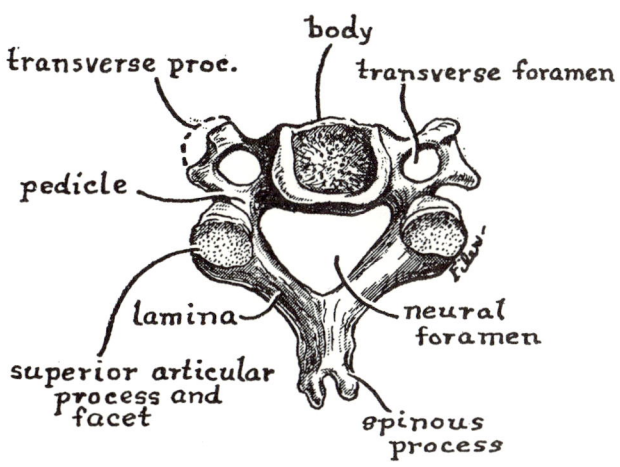

typical Cervical Vertebra

Figure 29.

ATYPICAL CERVICAL VERTEBRAE. (See Fig. 30.)

First cervical (Atlas)

Body (unpaired). Replaced by a small anterior tubercle.

Spinous process (unpaired). Replaced by a small posterior tubercle.

Transverse process (paired). Extremely wide lateral projection.

Superior articular process (paired). Articulates with the occipital condyle.

Articular facet (unpaired). On the dorsal surface of the anterior tubercle for articulation with the odontoid process of the second cervical vertebra.

Second cervical (Axis)

Body (unpaired). Enlarged into the odontoid process superiorly. The odontoid process articulates with the dorsal surface of the anterior tubercle of the atlas.

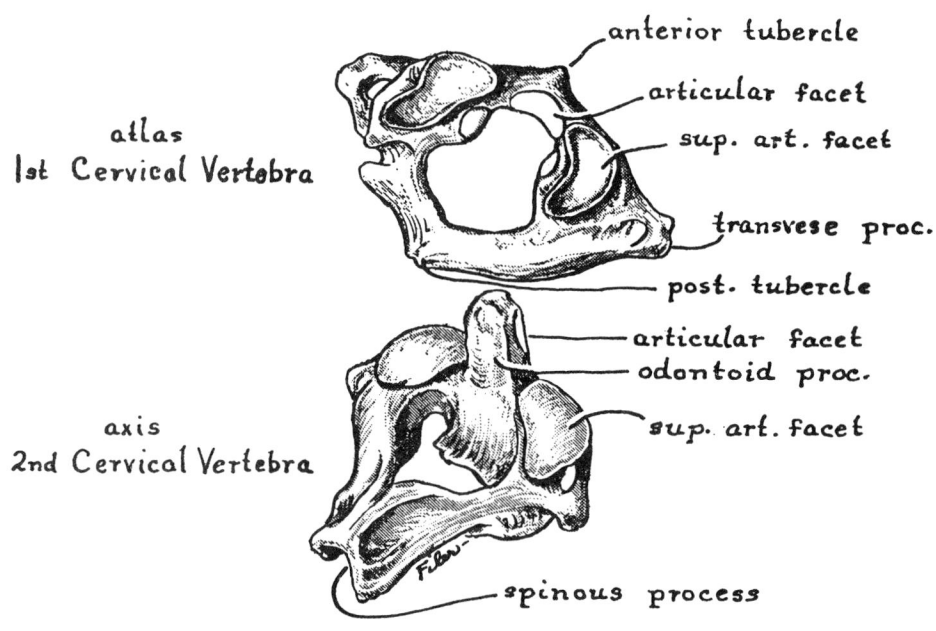

Figure 30.

TYPICAL THORACIC VERTEBRAE. (See Fig. 31.)

Body (unpaired). Typically has four demifacets, one on each dorsolateral corner, for articulation with the heads of the ribs.

Transverse process (paired). Has one facet on the anterior surface for articulation with the tubercle of a rib.

Spinous process (unpaired). Relatively long and tapering posterior projection.

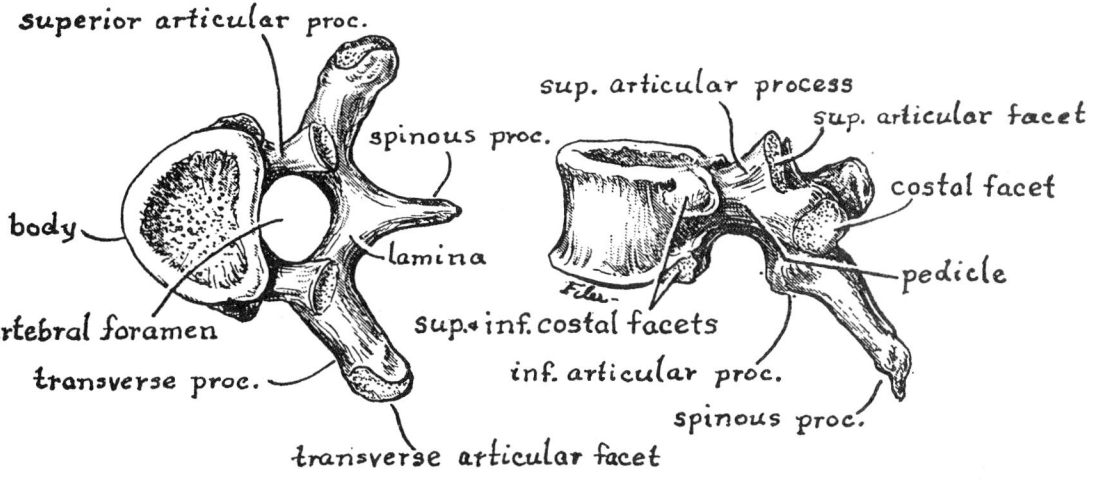

Figure 31.

ATYPICAL THORACIC VERTEBRAE

First thoracic

Body (unpaired). One whole facet on each side for articulation with rib one; one demifacet on each side for articulation with rib two.

Ninth thoracic

Body (unpaired). Has superior demifacets for rib nine but typically lacks inferior demifacets.

Tenth thoracic

Body (unpaired). Has single facets on each side for rib ten.

Eleventh and twelfth thoracic

Body (unpaired). Has single articular facets on each side for the corresponding ribs.

Transverse process (paired). Lacks articular facets.

LUMBAR VERTEBRAE. (See Fig. 32.)

Body (unpaired). Massive.

Process (paired). Blunt and massive.

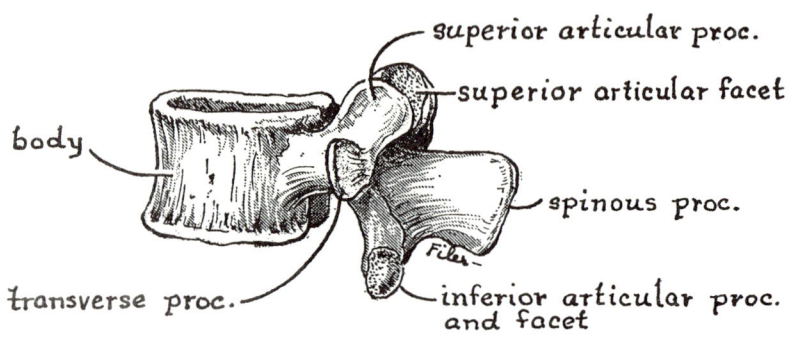

Lumbar Vertebra

Figure 32.

SACRAL VERTEBRAE. (See Fig. 33.)

Form. Lack usual form. Sacral vertebrae are fused into one piece, wedgeshaped, largest above, forming the back of the pelvis.

THE PELVIC GIRDLE

The pelvic girdle is composed of the paired **os inominatum**, a large irregular bone forming the anterior and lateral portion of the pelvis, and the **sacrum**, which completes the pelvis posteriorly. Each os inominatum is formed from three parts, distinct in the child but fused in the adult. These are the **ilium**, the broad, expanded upper portion; the **ischium**, the inferior and posterior portion; and the **pubis**, the anterior, inferior portion. The union of the three parts takes place in the **acetabulum**, a cup-shapd depression which receives the head of the femur. (See Figs. 34 and 35.)

ILIUM (paired).

Crest. The superior rim.

Anterior-superior spine. Anterior terminus of the crest.

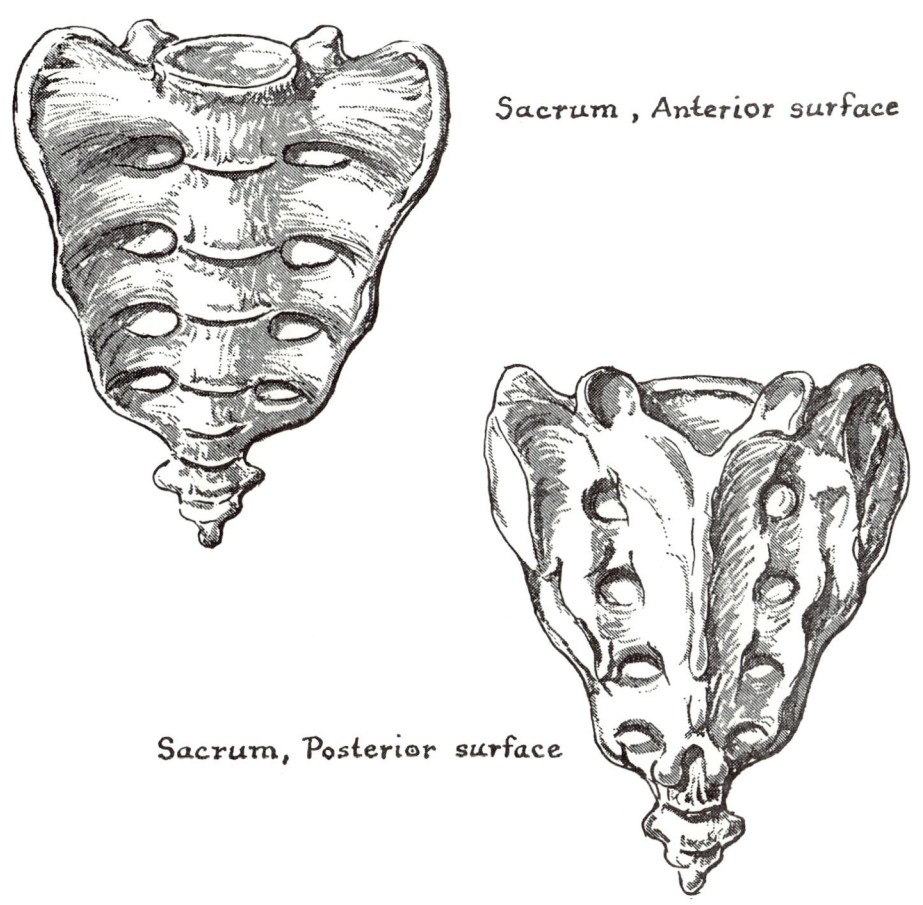

Figure 33.

Anterior-inferior spine. Projection below the anterior-superior spine.

Posterior-superior spine. Posterior terminus of the crest.

PUBIS (paired).

    Symphysis. Point of junction of the two pubic bones anteriorly at midline.

    Crest. Superior border lateral to the symphysis on each side.

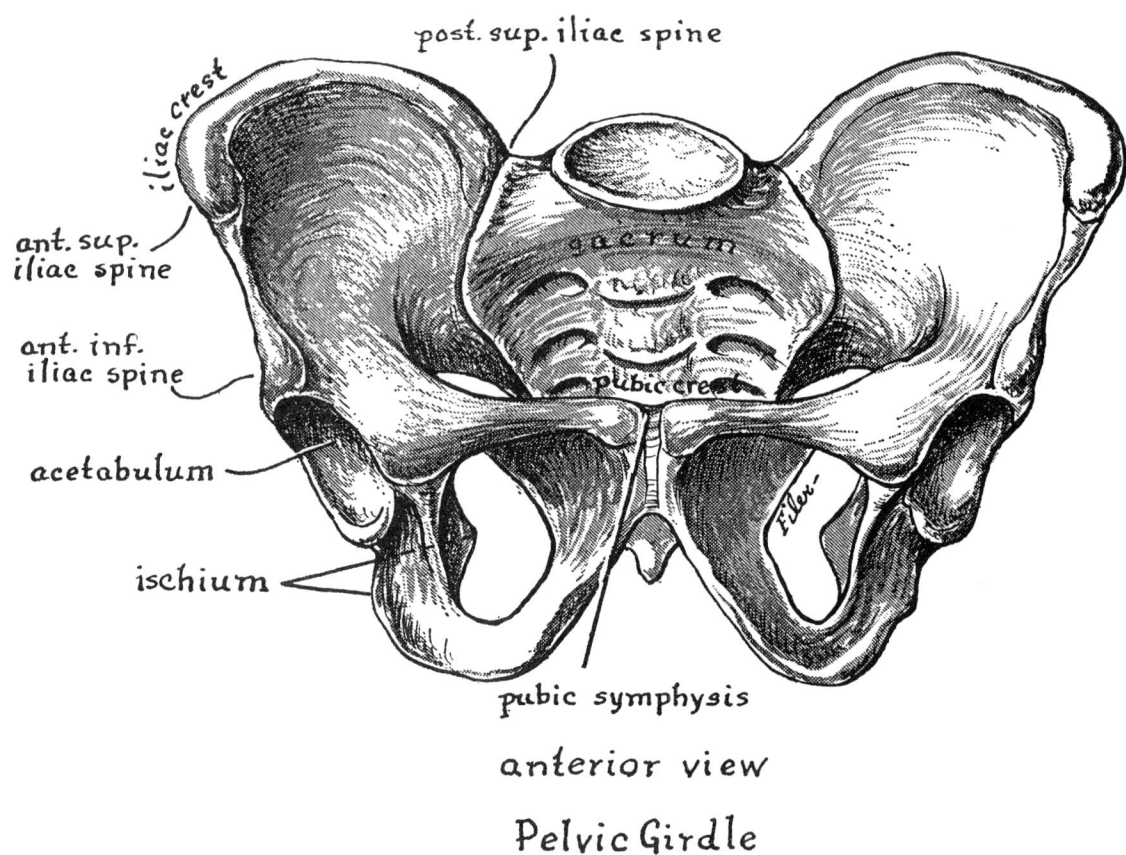

Figure 34.

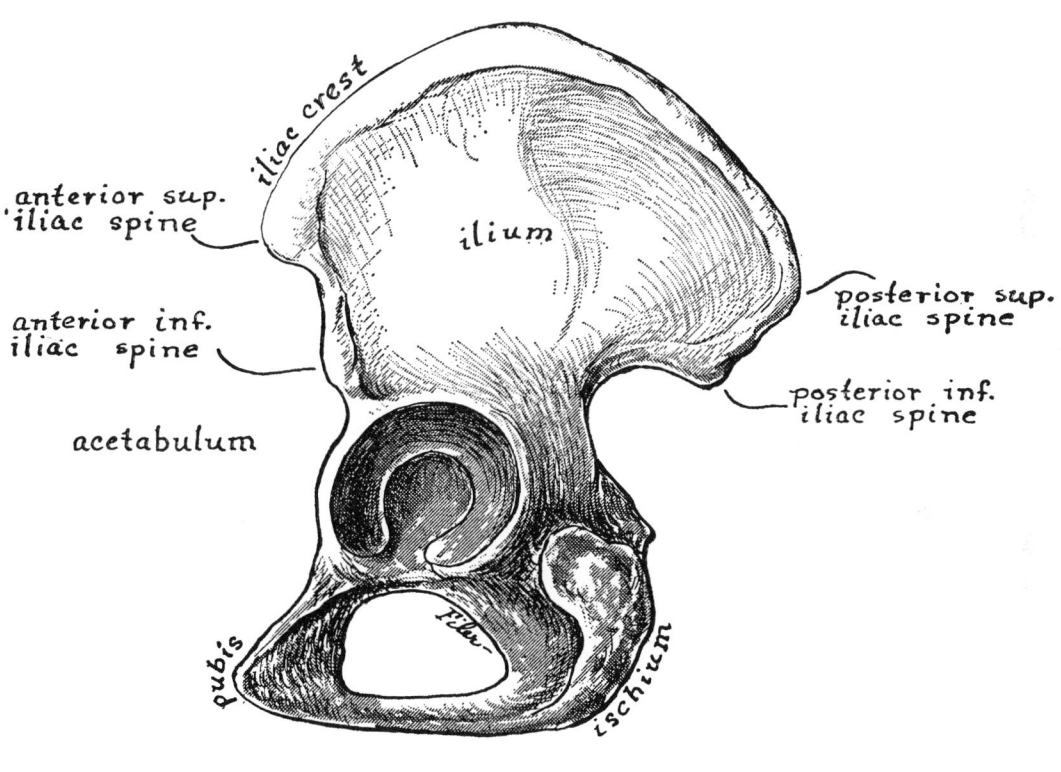

lateral view
Left Os Innominatum

Figure 35.

THE SHOULDER GIRDLE

The shoulder girdle is composed of the unpaired sternum anteriorly at midline, and the paired clavicle and scapula. (See Fig. 36.)

STERNUM (unpaired). (See Fig. 37)

Manubrium (unpaired). Roughly six-sided superior portion of the sternum. Articulates with the medial end of the clavicle and the cartilages or ribs one and two.

Corpus (unpaired). Flat blade of bone below and fused to the manubrium. Articulates with the cartilages of ribs two through ten.

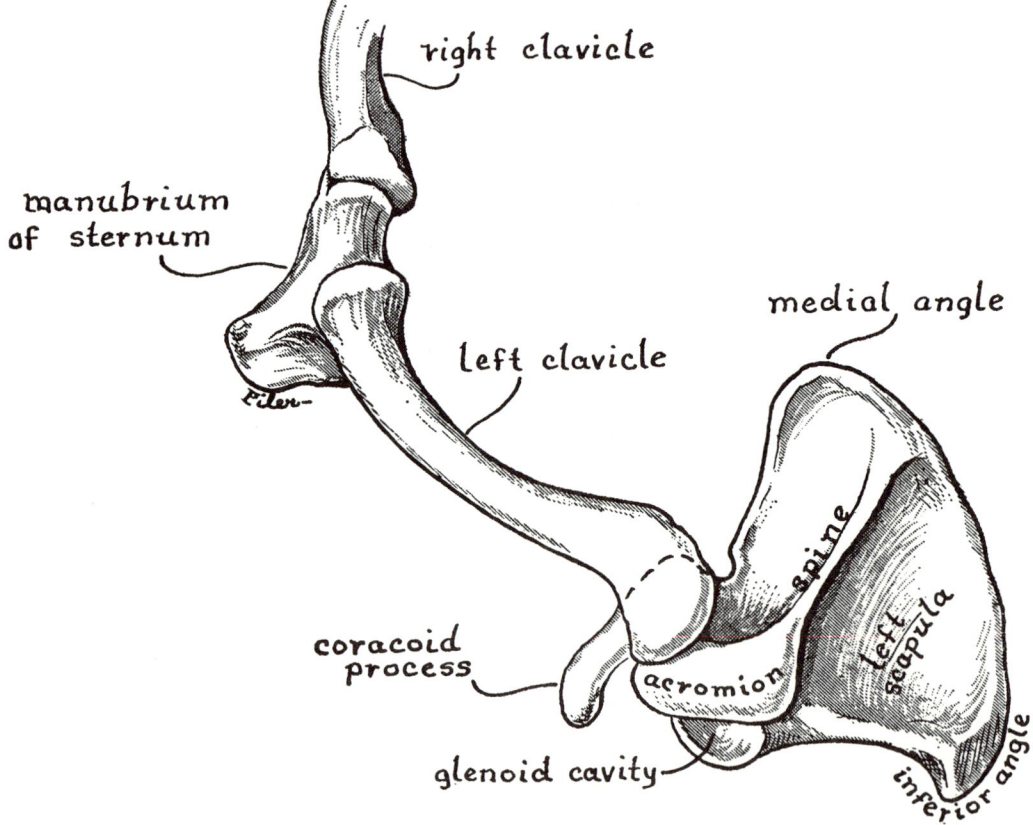

Figure 36.

Xiphoid process (unpaired). Small cartilagenous portion fused to the inferior end of the corpus.

CLAVICLE (paired). (See Fig. 38)

Body. Long curved bone. Articulates with the manubrium of the sternum medially and the acromion of the scapula laterally.

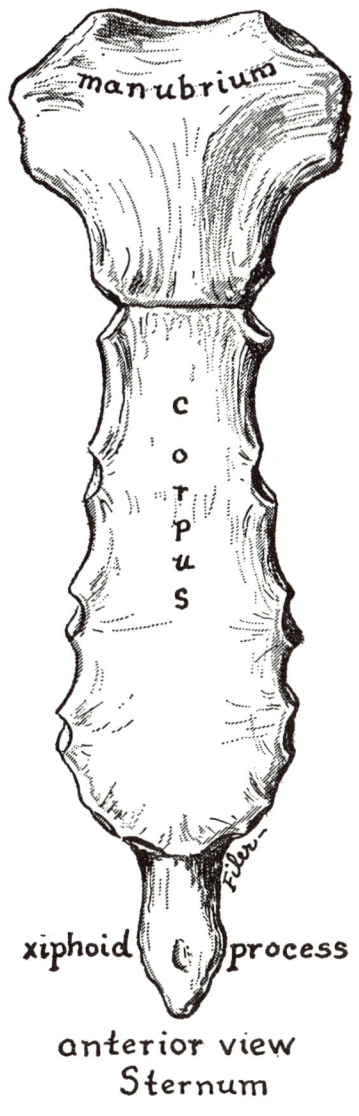

Figure 37.

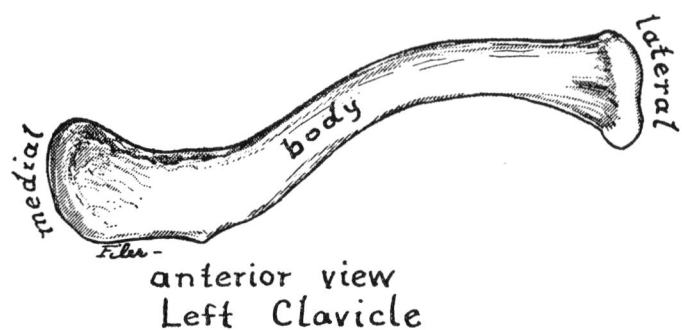

anterior view
Left Clavicle

Figure 38.

SCAPULA (paired). (See Fig. 39)

    Body. Roughly triangular in shape with a long vertical border lateral to the vertebral column. Superficial to the ribs in the back.

    Inferior angle. Inferior point of the body.

    Vertebral border. Medial border of the scapula lateral to the vertebral column.

    Medial angle. Superior terminus of the vertebral border.

    Spine. Crosses the dorsal surface obliquely to end laterally in the acromion.

    Acromion. Lateral terminus of the spine. Overhangs the superior end of the humerus.

    Glenoid cavity. Below the acromion. Articulation with the head of the humerus.

    Corocoid process. Extends anteriorly and slightly laterally from the superior border medial to the glenoid cavity.

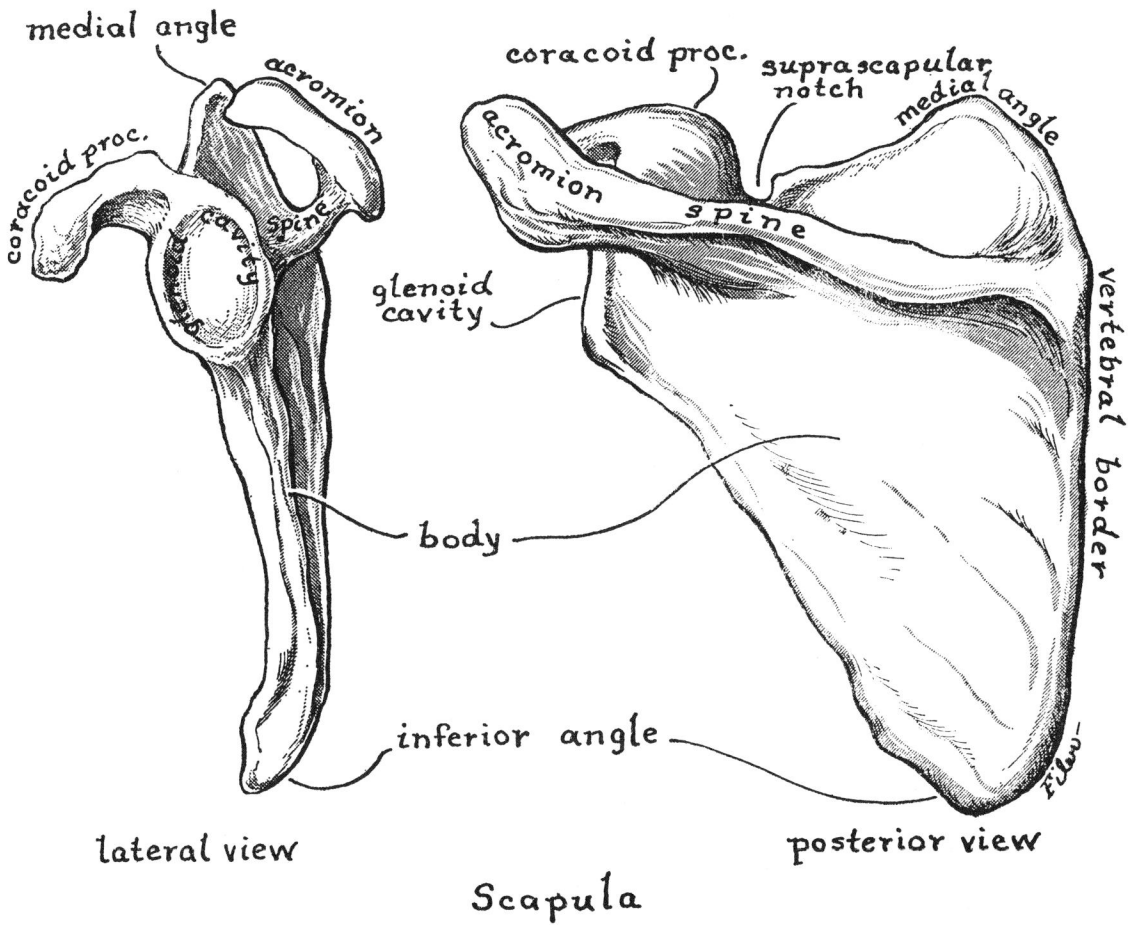

Figure 39.

THE RIBS

There are twelve pairs of ribs. All ribs articulate with the vertebrae posteriorly. Anteriorly, the upper six ribs articulate with the sternum by means of long costal cartilages. The seventh, eighth, ninth, and tenth ribs are joined to the sternum by a common costal cartilage. Ribs eleven and twelve have no anterior attachment and are called the floating ribs. The first rib is flat. Succeeding ribs become more complex in curvature. Length is increased through the seventh rib, obliquity through the ninth rib. Lower ribs are shorter and more horizontal. The eleventh and twelfth ribs do not articulate with the transverse processes of the vertebrae. (See Fig. 40.)

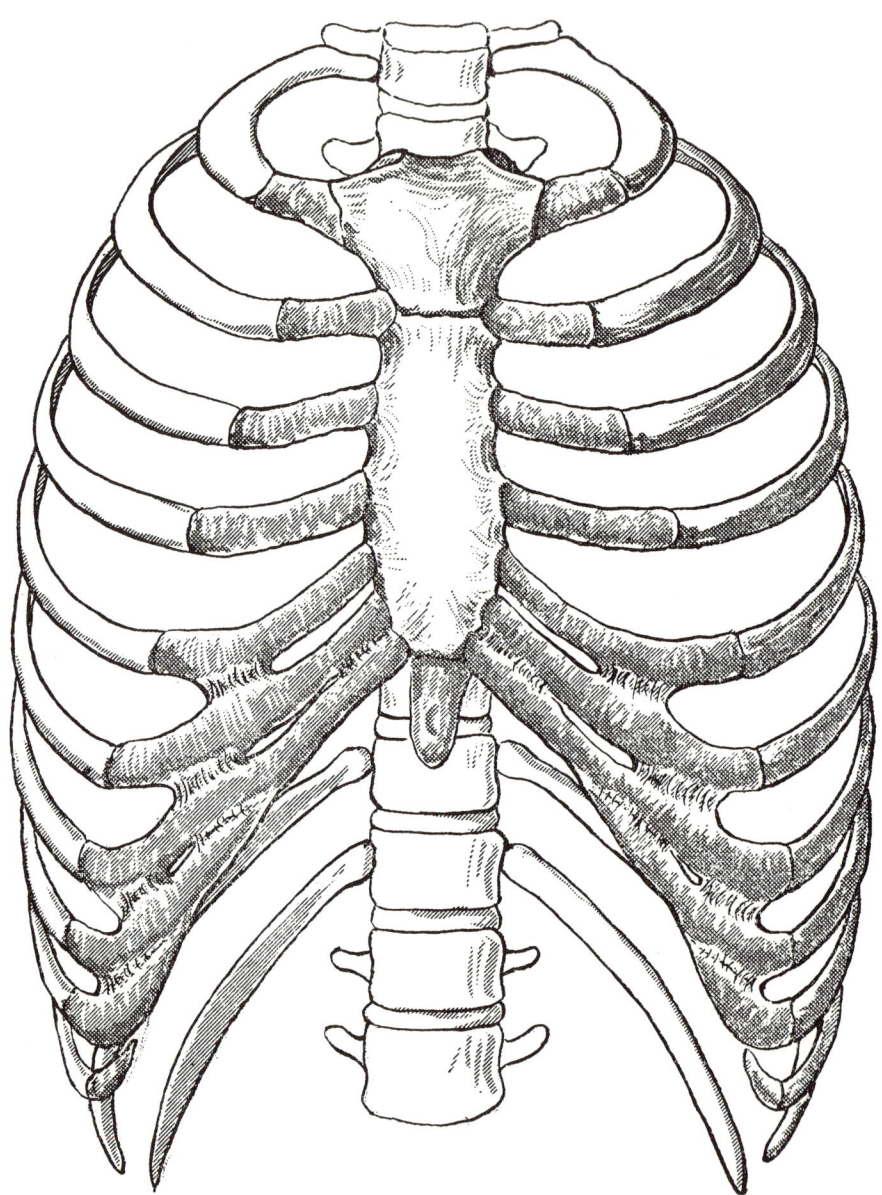

Thorax, Anterior view

Figure 40.

Skeletal Framework

TYPICAL RIB (paired). (See Figs. 41 and 42)

Head. At the medial end posteriorly. Articulates with the bodies of two adjacent vertebrae.

Neck. The short portion lateral to the head.

Tubercle. An elevation at the lateral end of the neck which articulates with the transverse process of a vertebra.

Angle. The most posterior part of the rib. The point where the rib turns from a dorsolateral direction to an anterolateral direction.

Shaft. The long part of the rib from the tubercle to the anterior end.

RIB MOVEMENT

Because of the complex curvature of the ribs, their movement on inhalation results in an increase in the anteroposterior, lateral, and vertical dimensions of the thorax. The lateral portions of the ribs

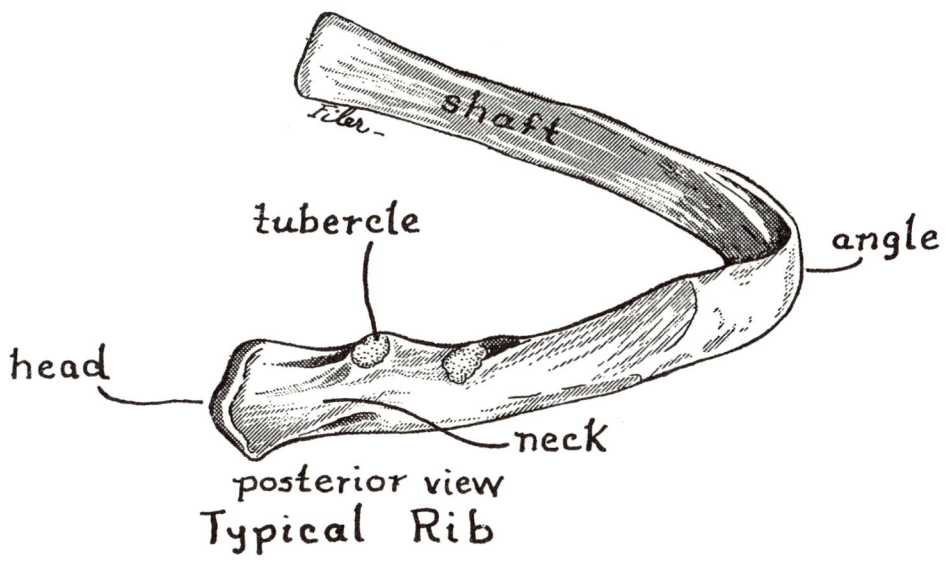

posterior view
Typical Rib

Figure 41.

rise in an upward and outward direction. The sternum moves upward and forward but maintains a constant angle to the body axis. All ribs move together but with more apparent elevation of the upper ribs and more expansion of the lower rib cage resulting from differences in rib shape from superior to inferior.

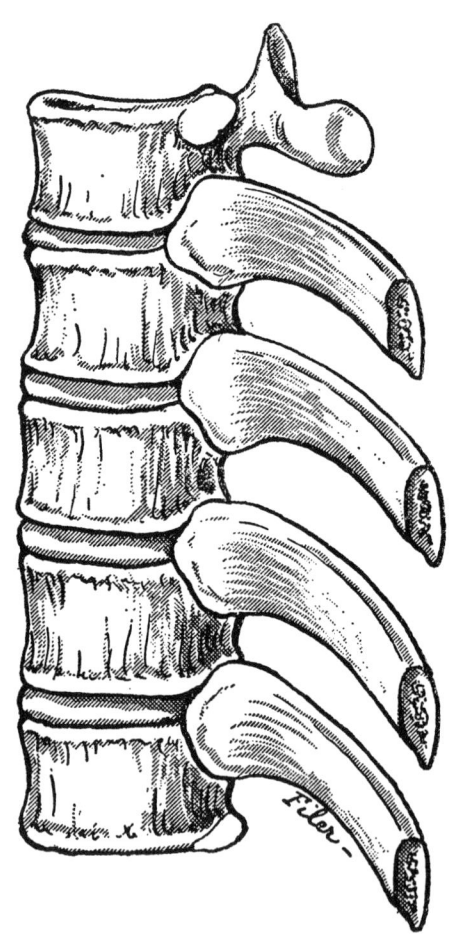

Typical Articulations of Ribs and Vertebrae

Figure 42.

III

RESPIRATION

Information regarding the muscles which produce the movements of inhalation and exhalation is incomplete. This is due in part to the fact that all muscles which may act during respiration also serve other body functions. Since most of these muscles probably contribute to movements of the arms, legs, head, and trunk, the muscles involved in respiration may vary from person to person and from moment to moment in the same person, depending on his posture and activity. It is therefore necessary to examine the muscles which *may* produce movements of respiration and look to their relative strength, direction of pull, and other functions in order to form an opinion on their relative importance to respiration. The muscles will be presented by area of the body in order to simplify their study in atlases and in dissection. Innervation of all of the muscles of respiration is presented at the end of the chapter so that the pattern of innervation may be better understood.

MUSCLES OF THE NECK

STERNOCLEIDOMASTOID (paired). Broad and thick. Lies superficially in the lateral portion of the neck. (See Fig. 43.)

ATTACHMENTS

To the lateral portion of the mastoid process of the temporal bone and to the adjacent portion of the superior nuchal line of the occipital bone.

To the anterior and superior surface of the manubrium of the sternum and by a separate head from the medial third of the clavicle.

FUNCTION

With one side acting, lateral flexion of the cervical vertebral column.

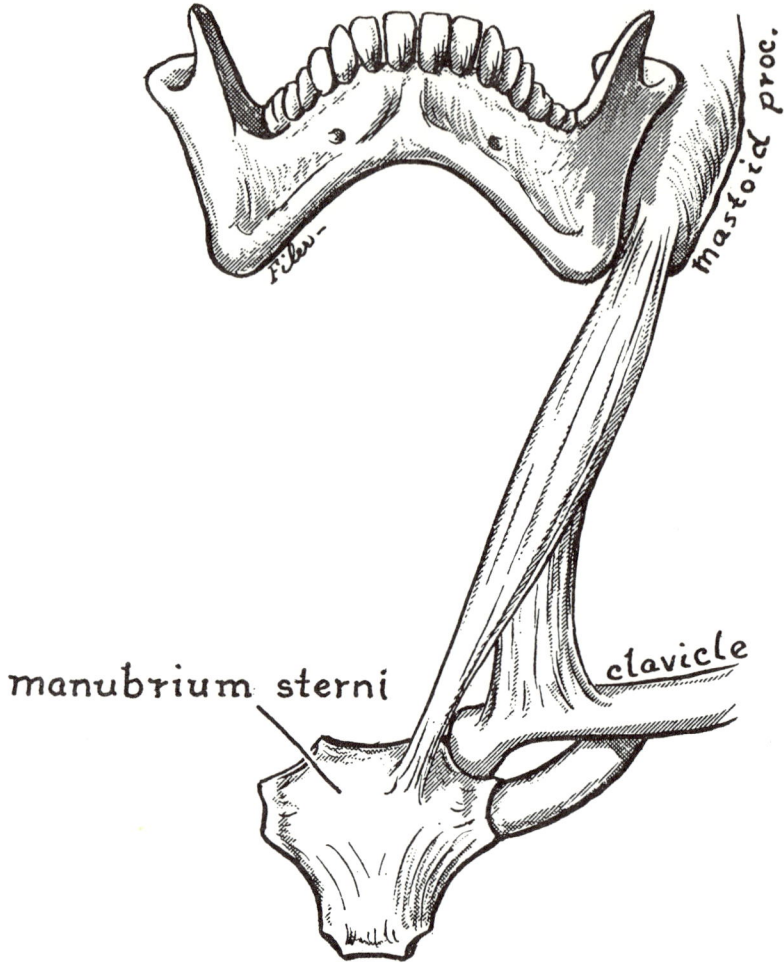

anterior view
Sternocleidomastoid Muscle

Figure 43.

With both sides acting, posterior flexion of the cervical vertebral column and upward rotation of the chin.

May help to raise the sternum in deep inhalation.

SCALENES (paired). Deep in the side of the neck. Well developed. Sometimes divided into three parts: anterior, medial, and posterior. (See Fig 44.)

Respiration

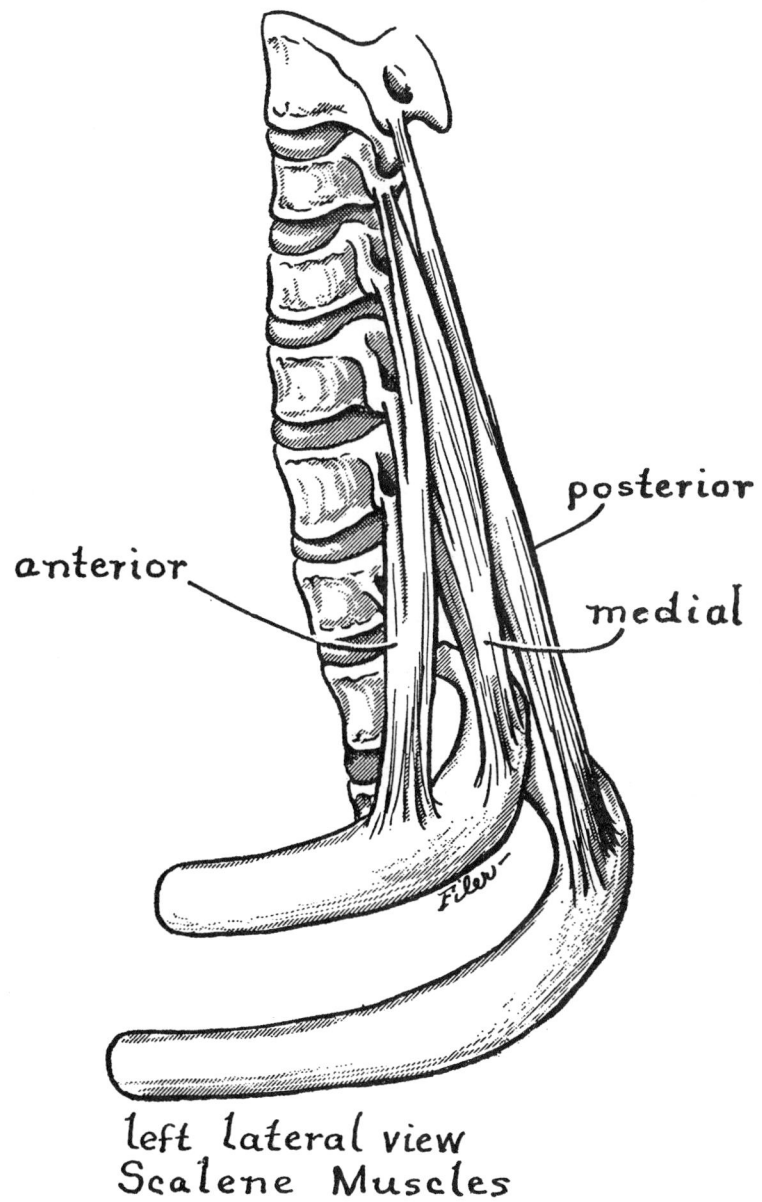

left lateral view
Scalene Muscles

Figure 44.

ATTACHMENTS

Anterior scalenes to the transverse processes of the third through sixth cervical vertebrae; *medial scalenes* to the transverse processes of the lower six cervical vertebrae; *posterior scalenes* to the transverse processes of the lower two or three cervical vertebrae.

Anterior and medial scalenes to the superior surface of the lateral portion of rib one; *posterior scalenes* to the superior surface of the lateral part of rib two.

FUNCTION

With one side acting, lateral flexion of the cervical vertebrae.

With both sides acting, may elevate ribs one and two in inhalation.

MUSCLES OF THE ANTERIOR THORAX

PECTORALIS MAJOR (paired). Broad and thick. Lies superficially on the anterior wall of the thorax. (See Fig. 45.)

ATTACHMENTS

To the medial half of the clavicle, anterior surface of the sternum, the cartilages of the upper six ribs, and the anterior layer of the abdominal aponeurosis.

All fibers converge to attach to the humerus below its upper head.

FUNCTION

Adducts and rotates the arm.

May aid in inhalation.

PECTORALIS MINOR (paired). Smaller than and deep to the pectoralis major. Fibers extend almost vertically. (See Fig. 46.)

ATTACHMENTS

To the corocoid process of the scapula.

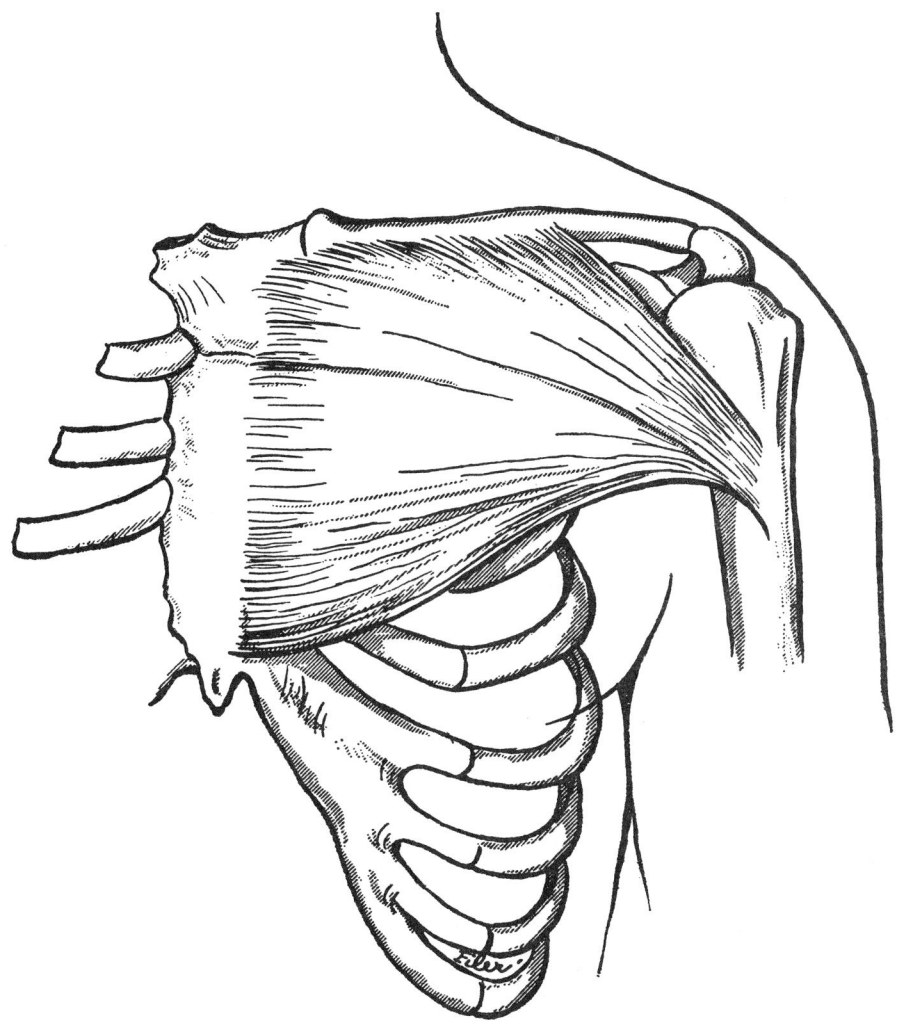

anterior view
Pectoralis Major Muscle

Figure 45.

To the anterior surfaces of ribs three through five just lateral to the ends of the ribs, and to the intercostal fascia.

FUNCTION

Draws the scapula downward and forward and may rotate it to aid in adduction of the arm.

May elevate ribs three through five in inhalation if the scapula is fixed.

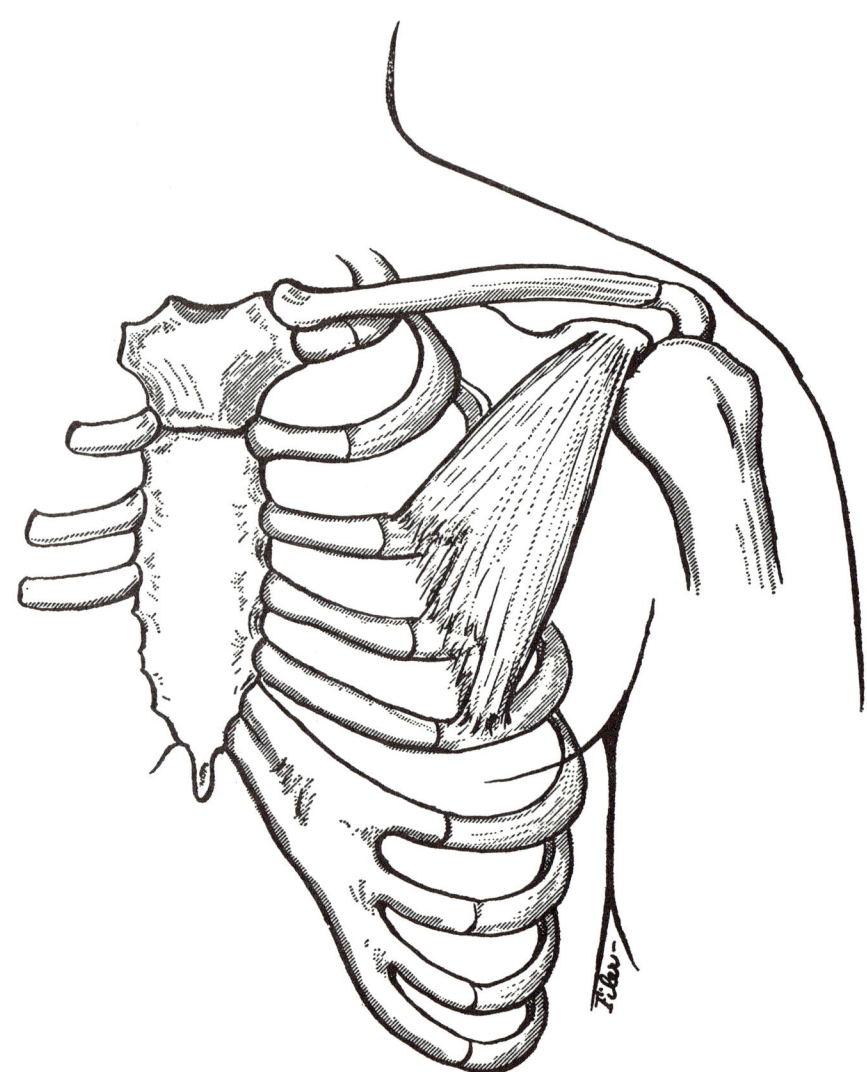

anterior view
Pectoralis Minor Muscle

Figure 46.

SUBCLAVIUS (paired). Small and slender. Lies between the clavicle and first rib. (See Fig. 47.)

ATTACHMENTS

To the junction of the first rib and its cartilage.

To the inferior surface of the clavicle above the corocoid process.

FUNCTION

May draw the shoulder downward and forward.

May exert a lifting force on rib one in inhalation if the clavicle is fixed.

SERRATUS ANTERIOR (paired). Broad and thick. Lies on the lateral and posterior thoracic wall superficial to the rib cage and deep to the scapula. (See Fig. 48.)

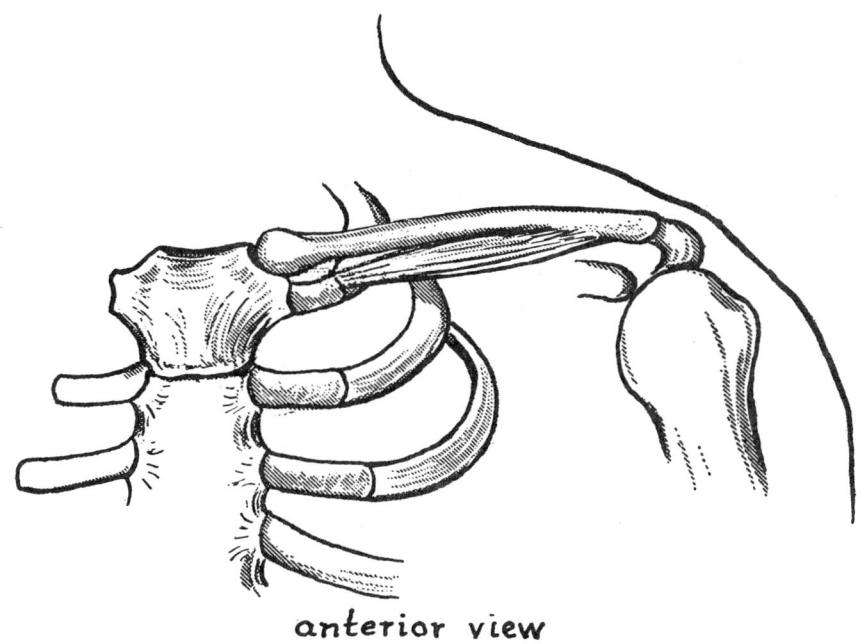

anterior view
Subclavius Muscle

Figure 47.

ATTACHMENTS

To the vertebral border of the scapula.

By digitations to the superficial surfaces of the upper eight or nine ribs and to the intercostal fascia.

FUNCTION

Rotates the scapula and draws it forward.

May aid in the lifting, expanding movement of the lateral rib cage in inhalation.

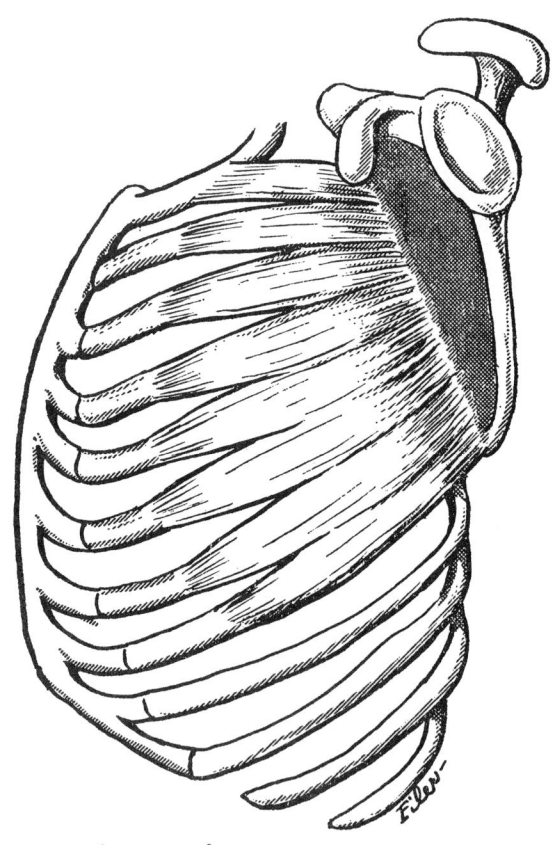

left lateral view
Serratus Anterior Muscle

Figure 48.

Respiration

EXTERNAL INTERCOSTALS (eleven pairs). Short and thin. Lie between the ribs. Course downward and medially on the anterior chest wall, downward and forward on the lateral wall, and downward and laterally on the posterior wall. (See Fig. 49).

ATTACHMENTS

To the lower edge of one rib and to the upper edge of the rib immediately below.

Extend from the tubercles of the ribs in the back to the junctions of the ribs and their cartilages in front.

FUNCTION

One side acting may aid in ipsilateral flexion of the trunk.

Both sides acting in cooperation with such muscles as the scalenes and pectoralis minor may aid in lifting the ribs for inhalation.

May simply act as binders to keep the ribs moving together in inhalation and exhalation.

INTERNAL INTERCOSTALS (eleven pairs). Thinner than and deep to the external intercostals. Course downward and laterally on the anterior chest wall, downward and back on the lateral wall, and downward and medially on the posterior wall. (See Fig. 49.)

ATTACHMENTS

To the lower edge of one rib and to the upper edge of the rib immediately below.

Extend from the rib angles in the back to the sternum in front.

FUNCTION

One side acting may aid in ipsilateral flexion of the trunk.

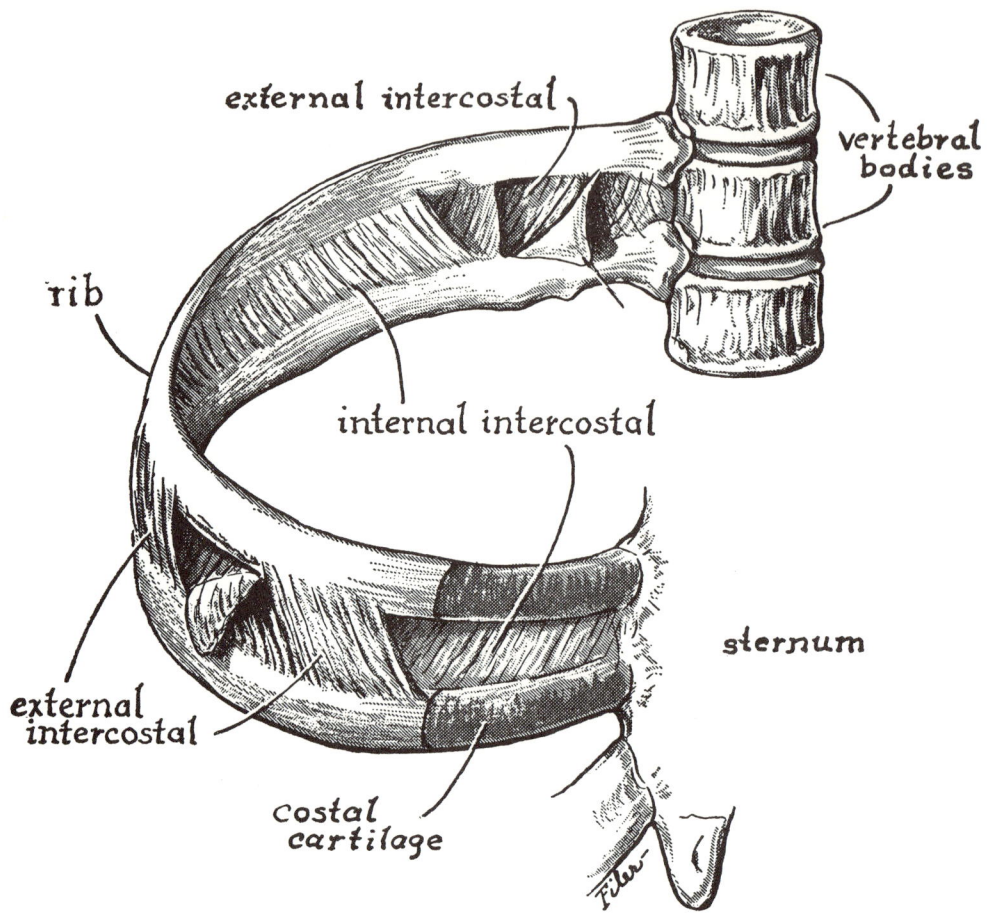

Intercostal Muscles

Figure 49.

Both sides acting in cooperation with such muscles as the external abdominal obliques may aid in lowering the ribs for exhalation.

May simply act as binders to keep the ribs moving together in inhalation and exhalation.

TRIANGULARIS STERNI (paired). Also called the **transversus thoracis.** Lies on the posterior surface of the ventral thoracic wall. Highly variable in form and extent from one person to another. (See Fig. 50.)

ATTACHMENTS

To the lower part of the deep surface of the sternum and adjacent costal cartilages.

To the deep surfaces of costal cartilages two through six.

FUNCTION

Because of its highly variable form and extent and its interlacing with tendonous fibers, its functional importance is doubtful.

May aid in lowering the ribs during exhalation.

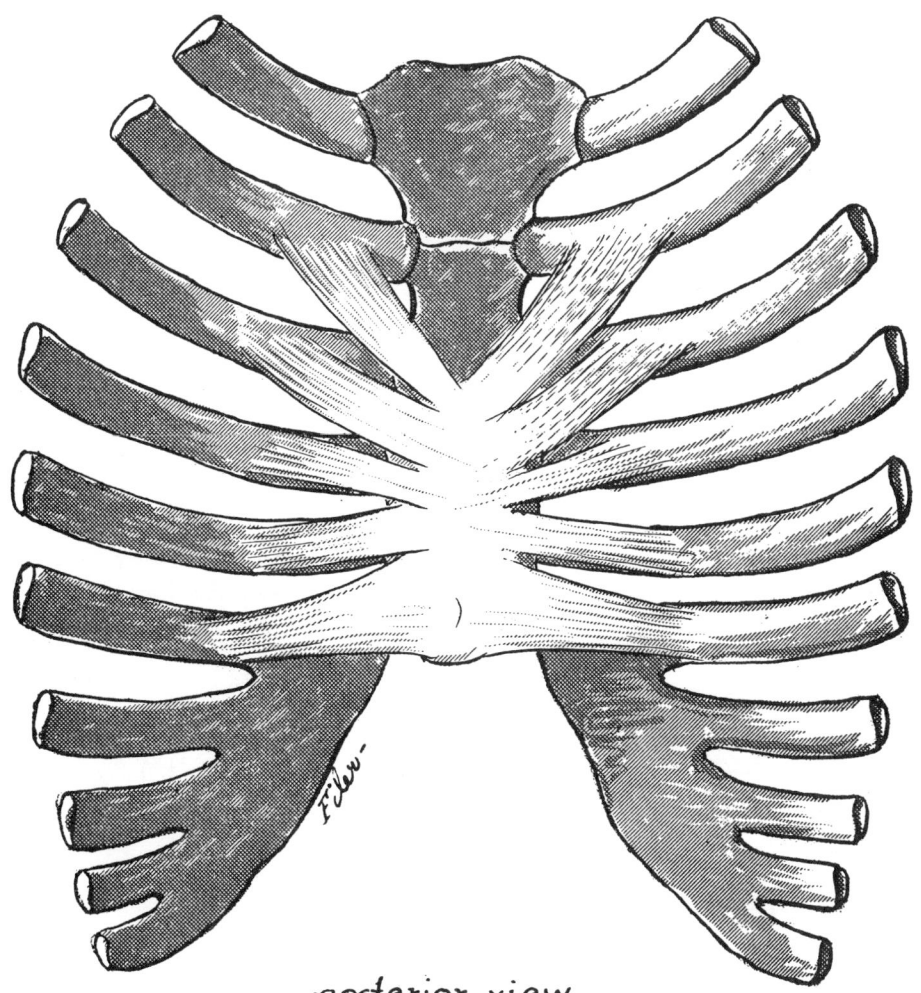

posterior view
Triangularis Sterni Muscle
Figure 50.

MUSCLES OF THE ABDOMEN

RECTUS ABDOMINIS (paired). Long and thick. Lies on either side of midline in the anterior abdominal wall. Courses vertically from the sternum to the pubis. In midline separating the two parts is a line of connective tissue, the **linea alba.** Lateral to the rectus and separating it from the lateral abdominal muscles is another line of connective tissue, the **linea semilunaris.** The rectus muscle is ensheathed in a layer of dense connective tissue, the **rectus sheath.** The linea semilunaris, rectus sheath, and linea alba together are called the **abdominal aponeurosis** formed from the broad sheets of tendon from the lateral abdominal muscles. (See Figs. 51 and 52.)

ATTACHMENTS

To the pubic crest.

To the superficial surfaces of the cartilages of ribs five through seven.

Divided into four parts by horizontal fibrous bands, the **tendinous intersections.**

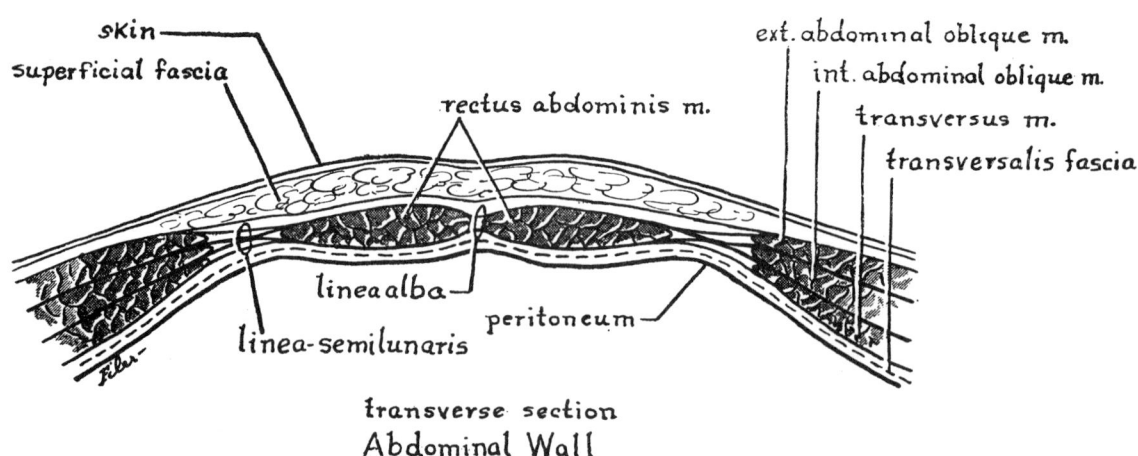

Figure 51.

FUNCTION

Anterior flexion of the vertebral column.

May aid in lowering the rib cage in exhalation.

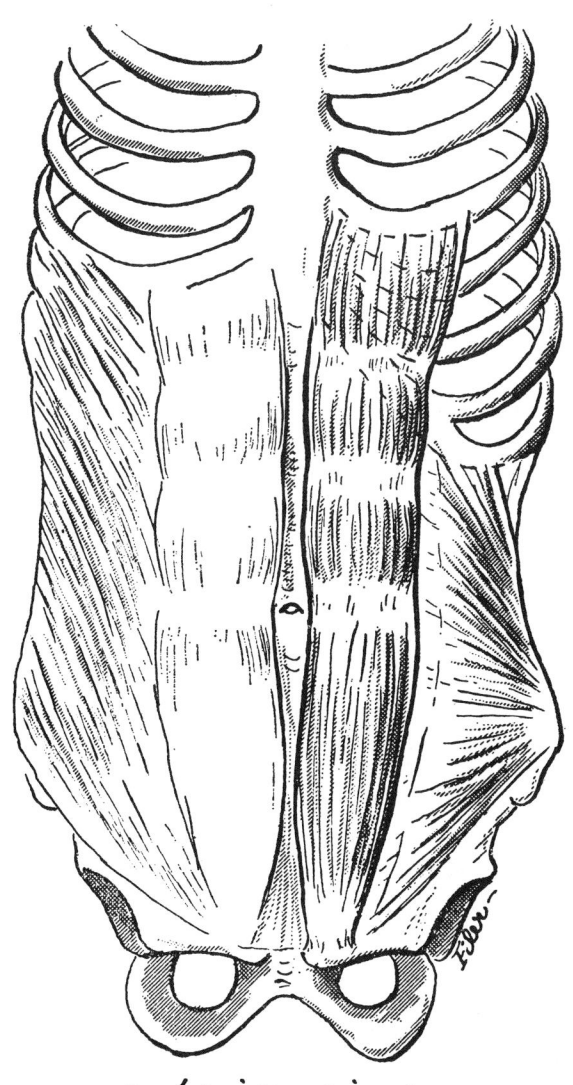

anterior view
Rectus Abdominis Muscle

Figure 52.

EXTERNAL ABDOMINAL OBLIQUE (paired). The thickest and most superficial sheet of muscle in the lateral abdominal wall. Courses downward and forward. (See Fig. 53.)

ATTACHMENTS

> To the external surfaces and inferior borders of the lower eight ribs. Interdigitates with the serratus anterior and the latissimus dorsi.
>
> To the anterior half of the iliac crest and via the superficial layer of the abdominal aponeurosis to the lateral half of the **inguinal ligament** (the lower thickened border of the external layer of the abdominal aponeurosis), the pubic crest, and the linea alba.

FUNCTION

> One side acting produces lateral flexion and rotation of the trunk. Both sides acting produce anterior flexion of the trunk and compression of the abdominal viscera.
>
> May lower the rib cage in exhalation and compress the abdominal viscera in opposition to the action of the diaphragm.

INTERNAL ABDOMINAL OBLIQUE (paired). Deep to the external abdominal oblique on the lateral abdominal wall. Courses downward and posteriorly. (See Fig. 54.)

ATTACHMENTS

> To the lateral half of the inguinal ligament to the anterior two-thirds of the iliac crest, and to the lower portion of the lumbar fascia.
>
> To the lower borders of the cartilages of the lowest three or four ribs, via the middle layer of the abdominal aponeurosis, to the linea alba, and to the pubic crest.

Respiration

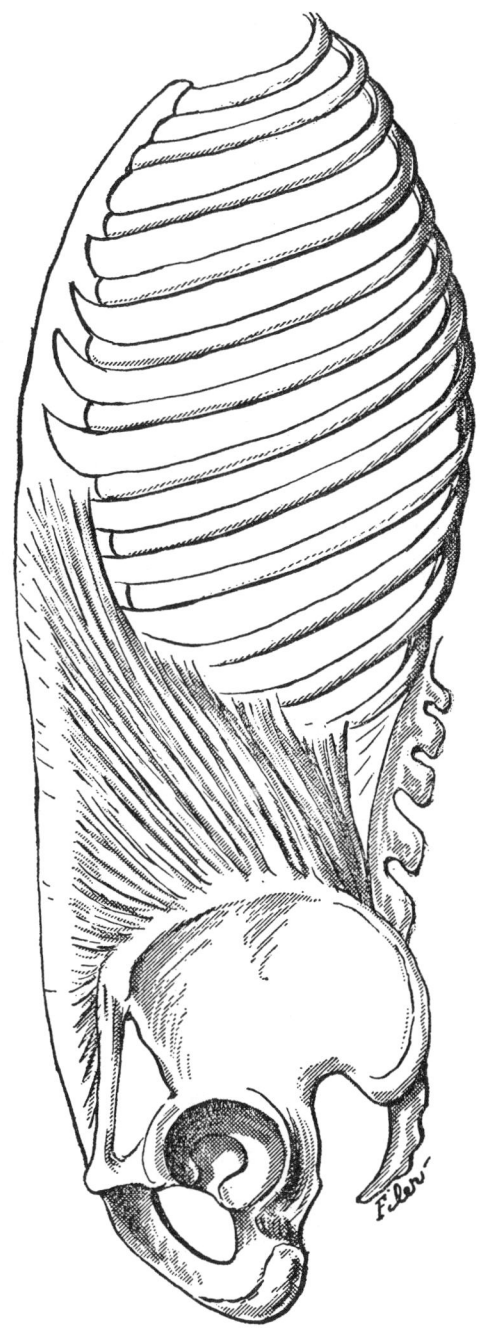

lateral view
Internal Oblique Muscle

Figure 53.

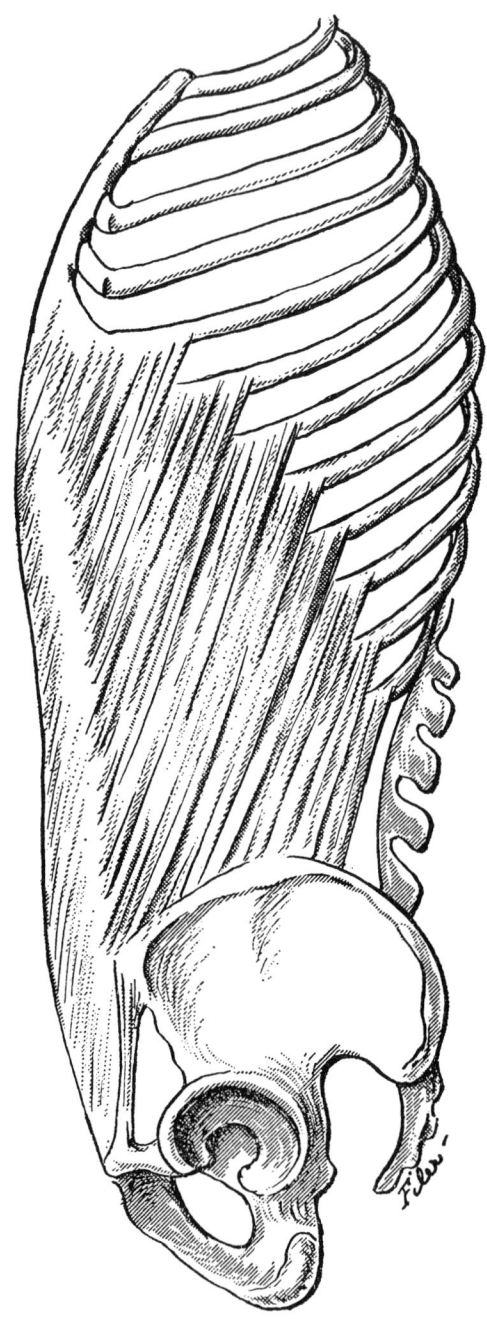

lateral view
External Oblique Muscle

Figure 54.

FUNCTION

One side acting produces lateral flexion and rotation of the trunk. Both sides acting produce anterior flexion of the trunk and compression of the abdominal viscera.

May lower the rib cage in exhalation and compress the abdominal viscera in opposition to the diaphragm.

TRANSVERSE ABDOMINAL (paired). Deepest and thinnest of the lateral abdominal muscles. Fibers course primarily in a transverse direction. (See Fig. 55.)

ATTACHMENTS

To the lateral third of the inguinal ligament, anterior three fourths of the iliac crest, lumbar fascia, and the deep surfaces of the cartilages of the lower six ribs. Interdigitates with the diaphragm.

To the pubic crest and via the deep layer of abdominal aponeurosis, to the linea alba.

FUNCTION

Compresses the abdominal viscera to act against the diaphragm in exhalation.

DIAPHRAGM (unpaired). Forms a partition separating the thoracic and abdominal cavities. Dome-shaped with a circular peripheral attachment to the inferior margin of the rib cage. Fibers rise and converge to a central tendon. The right side of the dome is slightly higher than the left. The diaphragm is penetrated by the esophagus, vena cava, and aorta. (See Figs. 56 and 57.)

ATTACHMENTS

Sternal part to the deep surface of the xiphoid process. *Costal part* to the anterior ends of ribs seven to eleven and their cartilages and to rib

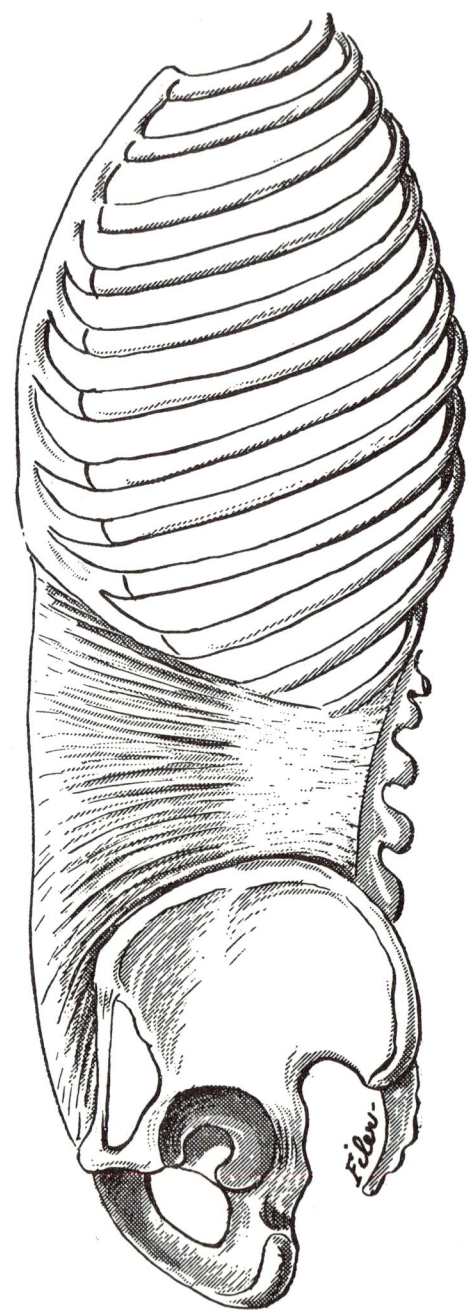

lateral view
Transverse Abdominal Muscle

Figure 55.

twelve. *Lumbar part* to the anterior surface of the upper four lumbar vertebrae.

To the central tendon.

FUNCTION

Draws the central tendon downward, compressing the abdominal viscera.

In lowering the central tendon in inhalation, increases the vertical dimension of the thorax.

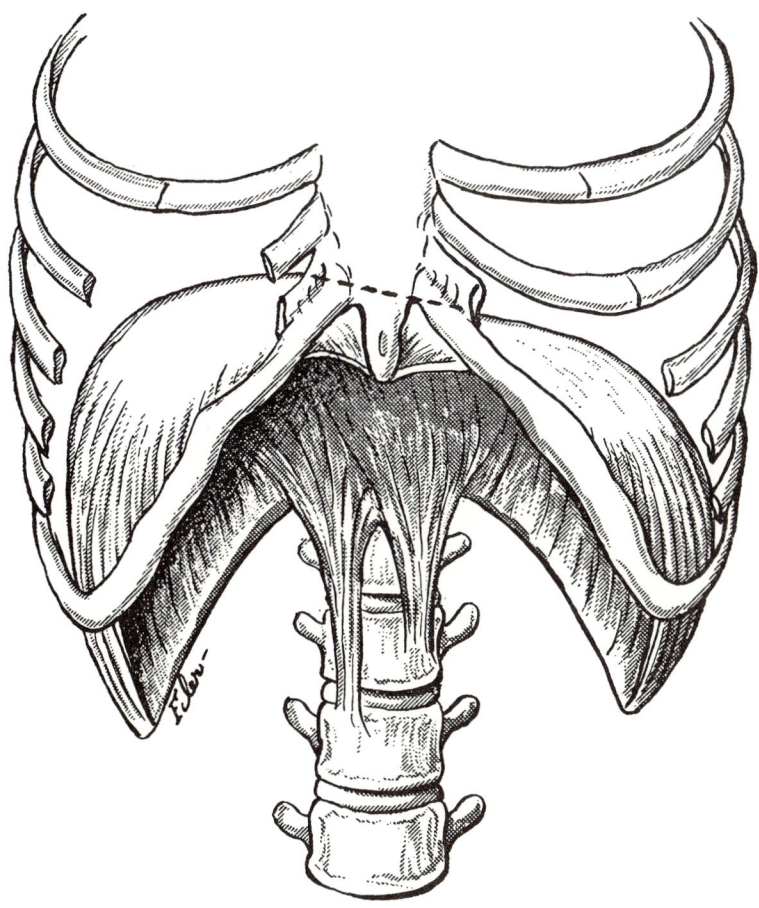

anterior view
Diaphragm

Figure 56.

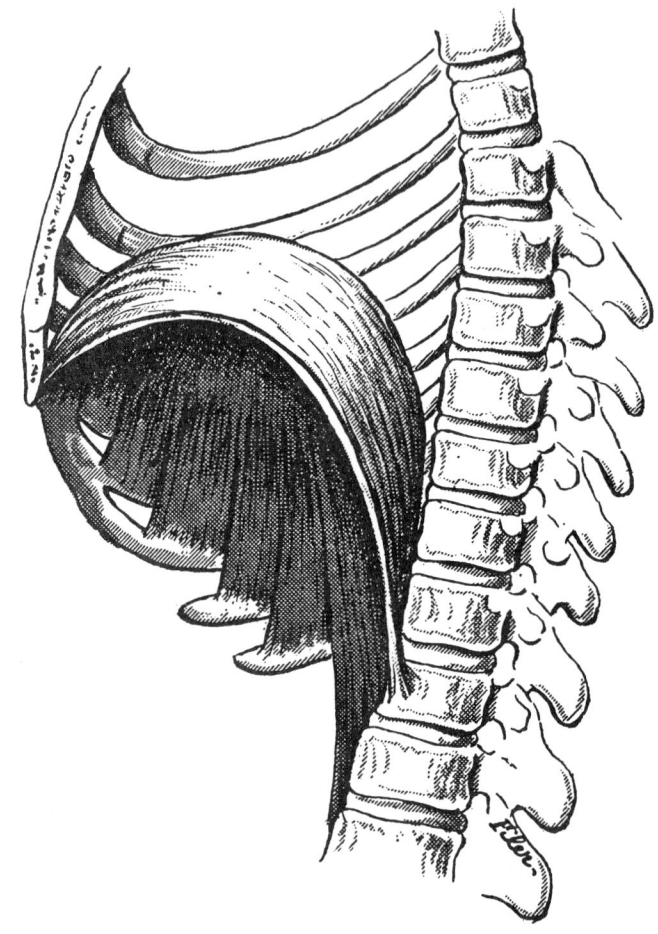

sagittal section
Diaphragm

Figure 57.

BACK MUSCLES

TRAPEZIUS (paired). Flat and triangular. Lies superficially in the upper back. (See Fig. 58.)

ATTACHMENTS

To the external occipital protuberance, the adjacent portion of the superior nuchal line, the nuchal ligament (covering and attaching to the spinous processes of the upper six cervical vertebrae), the spinous processes of the seventh cervical, and all thoracic vertebrae.

Fibers converge on the scapular spine, the acromion, and the lateral third of the clavicle.

FUNCTION

Rotates, raises, lowers, or adducts the scapula. Draws the head backward.

May brace the scapula against the pull of pectoralis minor during inhalation.

LATISSIMUS DORSI (paired). Large and thick. Lies superficially in the lower back, deep only to the portion of the trapezius attaching to the lower thoracic vertebrae. (See Fig. 59.)

ATTACHMENTS

To the spinous processes of the lower six thoracic and all lumbar and sacral vertebrae, and to the lumbar fascia. Some fibers attach to the superficial surface of the lower three or four ribs and to the iliac crest.

All fibers converge on the humerus just below its superior head.

FUNCTION

Adducts and rotates the arm. Draws the shoulder downward and backward.

Fibers attaching to the lower ribs may aid in their elevation in inhalation or in lowering the ribs in exhalation.

SERRATUS POSTERIOR SUPERIOR (paired). An extremely thin short muscle with a long tendon. Deep to the trapezius. (See Fig. 60.)

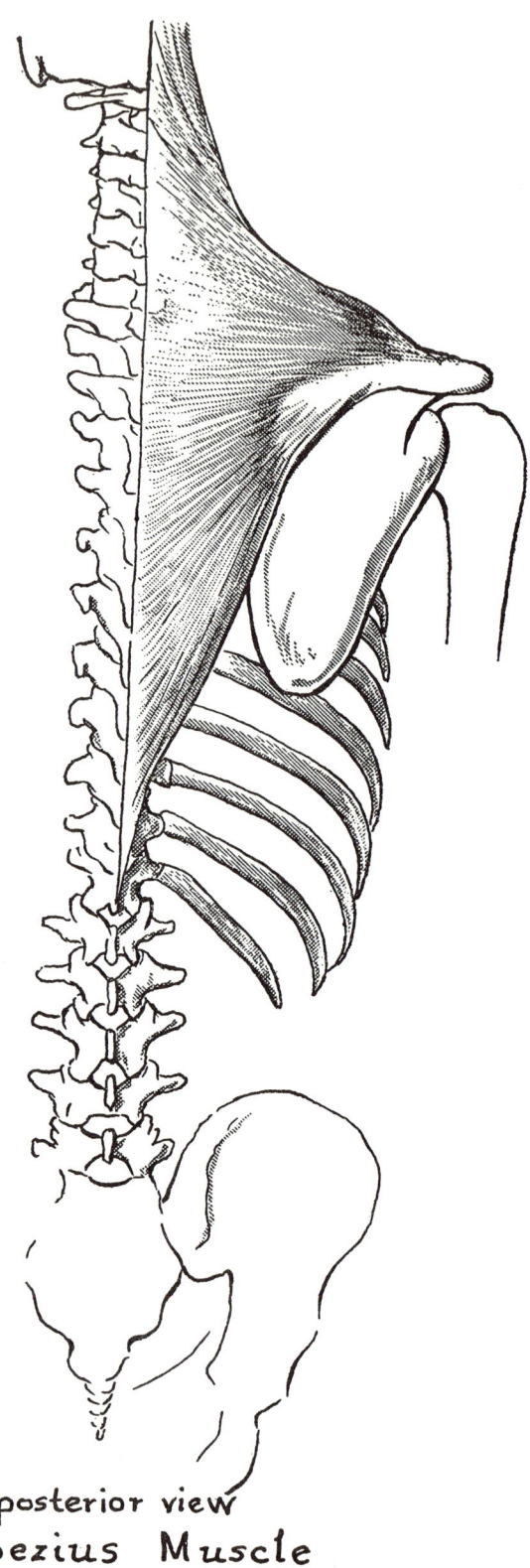

posterior view
Trapezius Muscle

Figure 58.

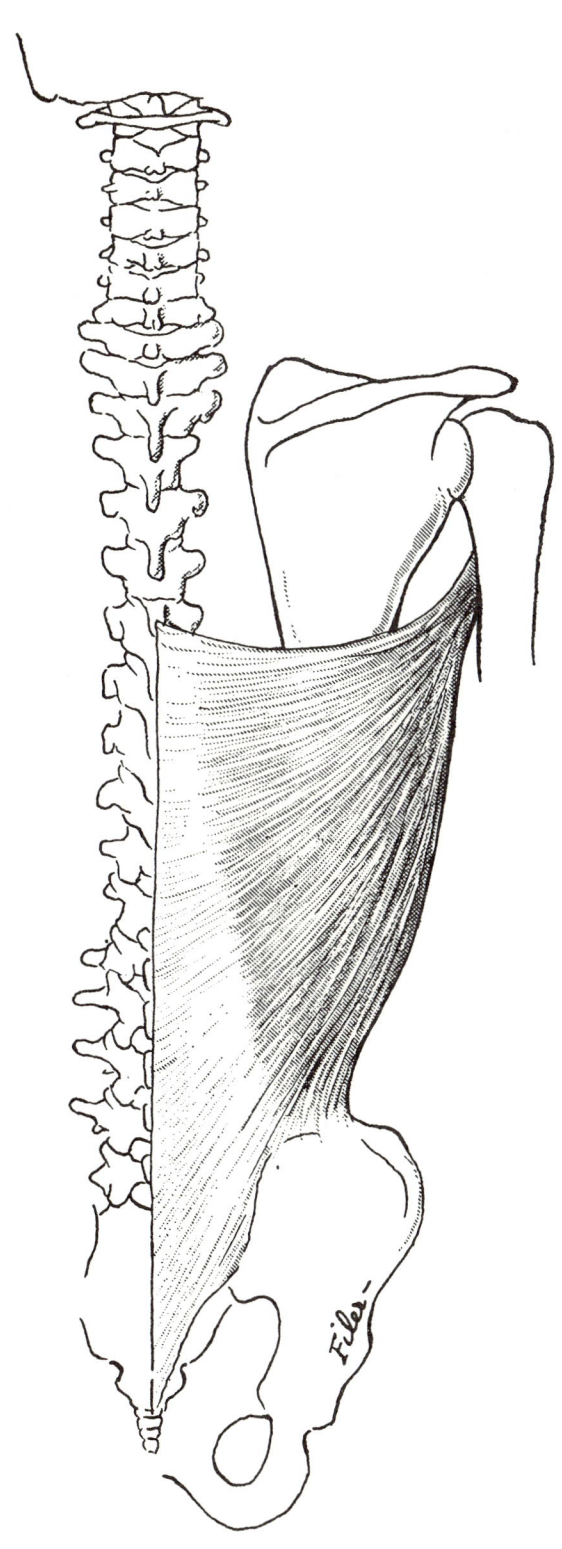

posterior view
Latissimus Dorsi Muscle

Figure 59.

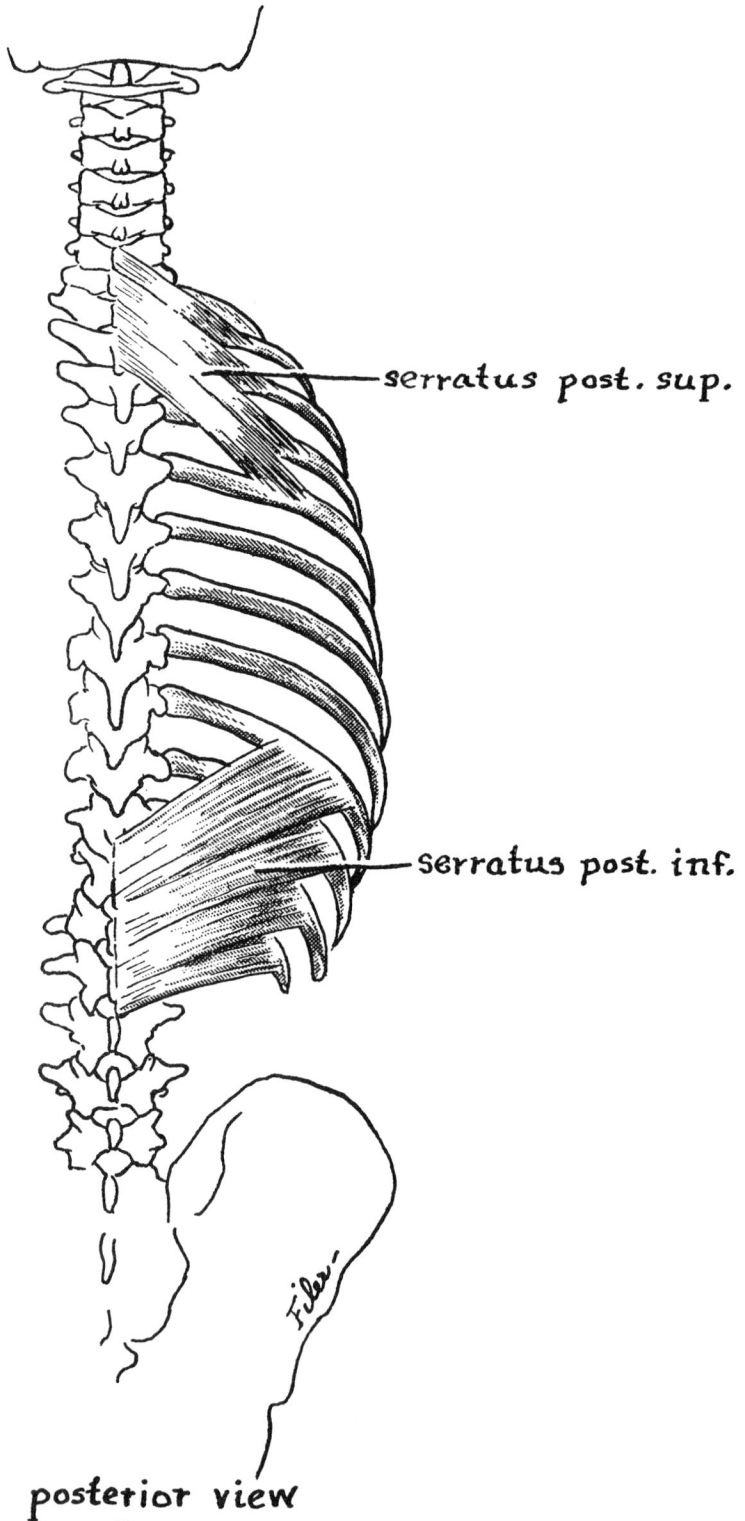

posterior view
Serratus Posterior Muscles

Figure 60.

ATTACHMENTS

By a long tendon to the nuchal ligament and the spinous processes of the seventh cervical and upper two or three thoracic vertebrae.

To the upper borders of ribs two through five just beyond the angles of the ribs.

FUNCTION

One side acting may aid in lateral flexion of the trunk.

Both sides acting may aid in elevation of the ribs in inhalation.

SERRATUS POSTERIOR INFERIOR (paired). Somewhat more muscular than the serratus posterior superior. Lies deep to the latissimus dorsi.

ATTACHMENTS

By a long tendon to the spinous processes of the lowest two or three thoracic and upper two or three lumbar vertebrae.

To the inferior borders of the lowest four ribs just lateral to their angles.

FUNCTION

One side acting may aid in lateral flexion of the trunk.

Both sides acting may aid in lowering the ribs during exhalation.

ILIOCOSTALIS CERVICIS (paired). Part of the **sacrospinal** muscle group. (See Fig. 61.)

ATTACHMENTS

To the transverse processes of the fourth, fifth, and sixth cervical vertebrae.

To the superficial surfaces of the third through sixth ribs at their angles.

FUNCTION

One side acting may aid in lateral flexion of the upper vertebral column. Both sides acting extend the vertebral column.

May aid in elevation of the ribs during inhalation.

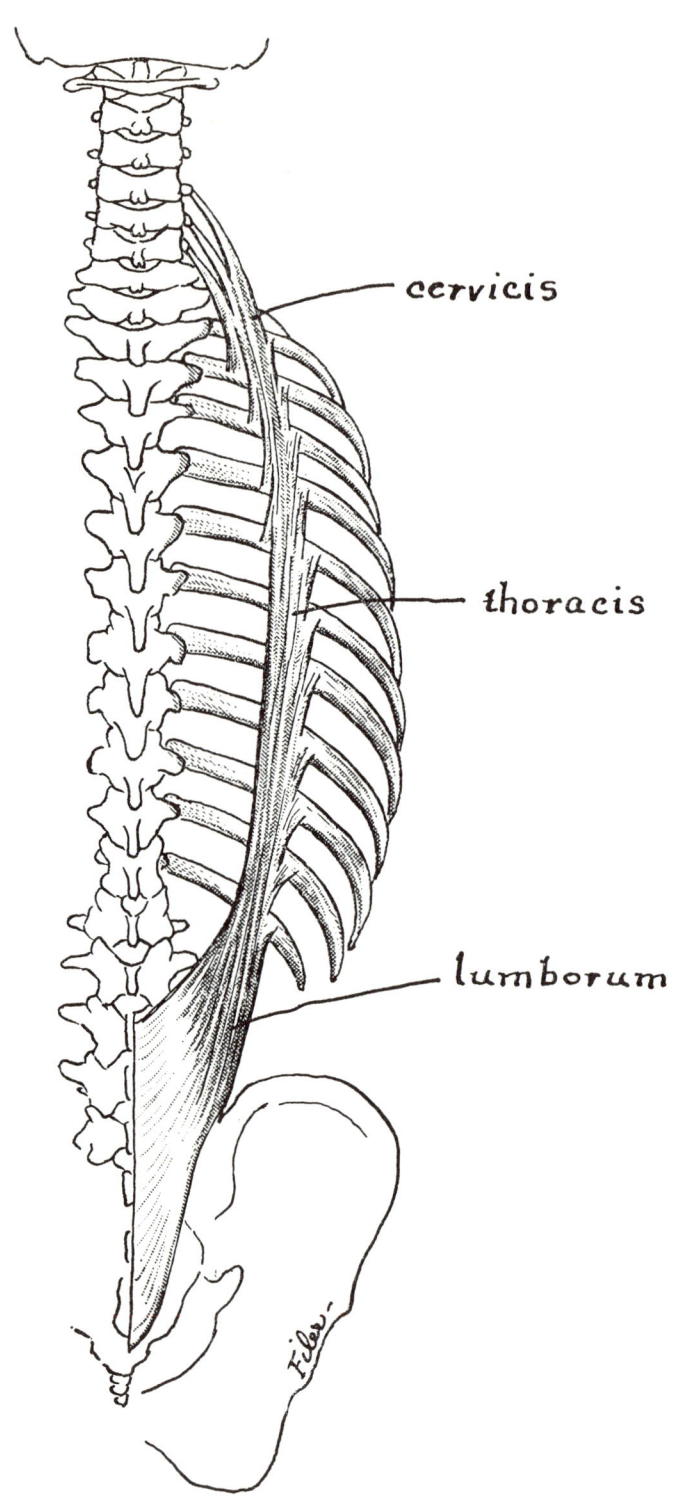

posterior view
Ilio-costalis Muscles

Figure 61.

ILIOCOSTALIS THORACIS (paired). Also called the **iliocostalis dorsi.**

Part of the **sacrospinalis** muscle group.

ATTACHMENTS

To the upper borders of the lower six ribs at their angles.

To the upper borders of the upper six ribs at their angles.

FUNCTION

One side acting may aid in lateral flexion of the vertebral column.

Both sides acting extend the vertebral column.

May aid in movements of the ribs during respiration.

ILIOCOSTALIS LUMBORUM (paired). Part of the **sacrospinal** muscle group.

ATTACHMENTS

To the lower borders of the lower six ribs at their angles.

To the lumbar fascia and the posterior part of the iliac crest.

FUNCTION

One side acting may aid in lateral flexion of the vertebral column. Both sides acting extend the vertebral column.

May aid in lowering the ribs during exhalation.

COSTAL ELEVATORS. (Twelve pairs). Lie lateral to the vertebral column on each side. (See Fig. 62.)

ATTACHMENTS

To the transverse processes of the seventh cervical and upper eleven thoracic vertebrae.

To the outer surface of the rib immediately below, between the tubercle and angle. Some of the lower parts of this muscle pass over one rib to attach two ribs below their upper attachment.

FUNCTION

One side acting will flex the vertebral column laterally. Both sides acting will extend the vertebral column.

May aid in rib elevation during inhalation.

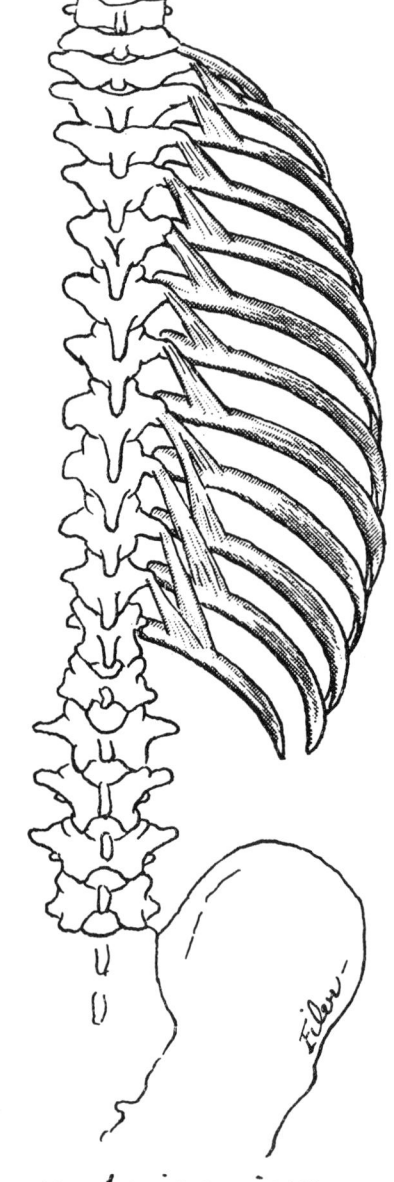

posterior view
Costal Elevators

Figure 62.

Respiration 81

QUADRATUS LUMBORUM (paired). Quadrilateral. Lies between the twelfth rib and the pelvis lateral to the vertebral column. (See Fig. 63.)

ATTACHMENTS

To the posterior half of the iliac crest and the transverse processes of the upper four lumbar vertebrae.

To the posterior half of the twelfth rib.

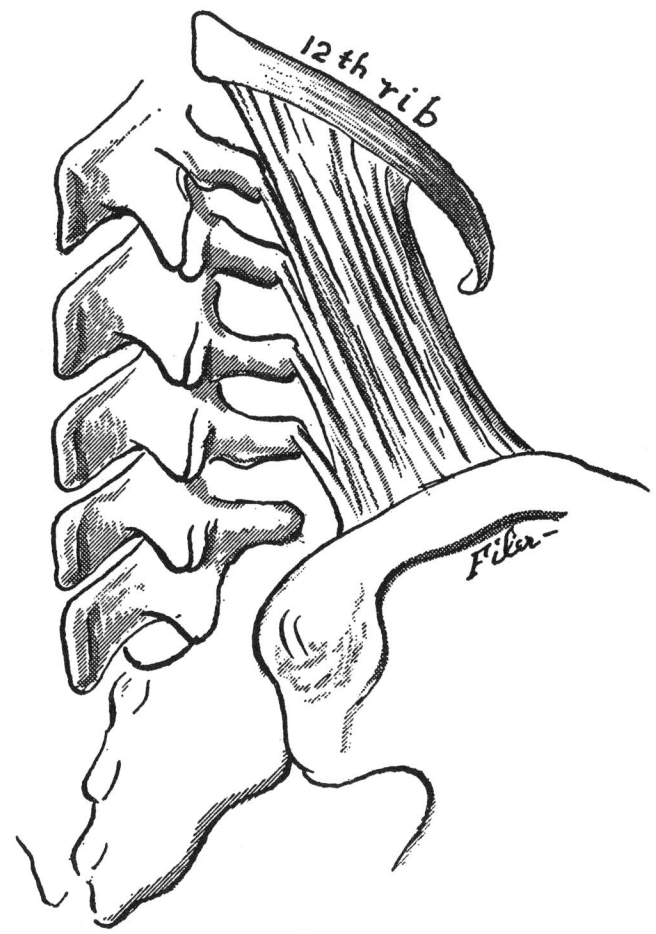

posterior lateral view
Quadratus Lumborum Muscle

Figure 63.

FUNCTION

One side acting flexes the vertebral column laterally.

May draw the last rib downward in exhalation.

SUBCOSTALS (paired). Inconsistent in form and extent. Lie on the anterior surface of the dorsal thoracic wall. Usually only well developed in the inferior thorax. (See Fig. 64.)

ATTACHMENTS

To the deep surfaces of the ribs near their angles.

To the deep surface of the second or third rib below, near the vertebral column.

FUNCTION

One side acting may aid in lateral flexion of the trunk.

May aid in lowering the ribs during exhalation.

THE TRACHEA AND LUNGS

The trachea is an open tube which extends inferiorly from the larynx in the anterior part of the neck. Its fibrous walls are held open by horseshoe-shaped cartilages which are found throughout its length. Inferiorly, about the level of the fifth thoracic vertebra, it divides into left and right bronchi, each of which supplies one lung. Each bronchus in turn divides and continues to subdivide until each ends in a number of alveoli. In the alveoli the gaseous exchanges of respiration take place. The total epithelial area in the adult lung during a deep inhalation has been estimated at nearly 230 square feet.

The lung tissue is passive and responds to the changing air pressure in the thoracic space as the rib cage and diaphragm move during inhalation and exhalation.

PERIPHERAL INNERVATION OF THE MUSCLES OF RESPIRATION

The spinal portion of the eleventh cranial nerve (spinal accessory) and the cervical, thoracic, and lumbar spinal nerves supply the muscles of respiration. Fibers from various spinal nerves join to form complex interconnections called **plexes.** Cervical nerves one through four, which arise from

Respiration

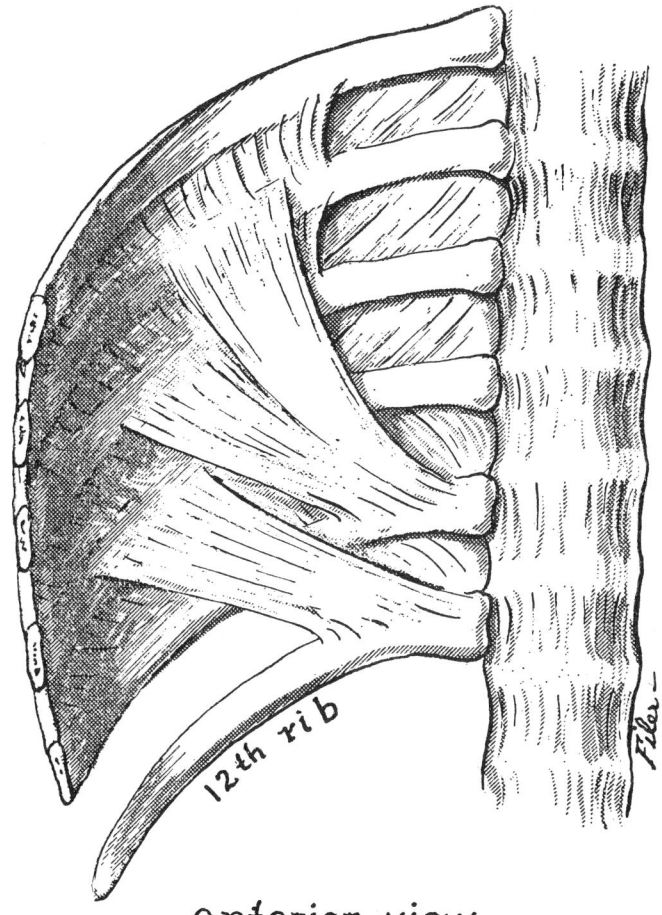

anterior view

Right Subcostal Muscles

Figure 64.

the spinal cord at the level of the first four cervical vertebrae, form the **cervical plexus**, which sends motor nerves to the sternocleidomastoid, trapezius, scalenes, and diaphragm. Cervical nerves five through eight and the first thoracic nerve which exits from the spinal cord between the corresponding vertebrae form the **brachial plexus**, which sends motor fibers to the pectoralis major, pectoralis minor, serratus anterior, and latissimus dorsi. The **lumbar plexus** is formed from fibers from the first four lumbar nerves and sometimes from fibers of the last thoracic nerve. Fibers from the lumbar plexus supply the transverse abdominal muscle.

Other muscles and some of those named above are supplied by spinal nerves which exit from the spinal cord at the level of the muscle. The distribution of all of these nerves is not clear and is variable from one person to another. The following list of efferent innervation represents what seems to be the consensus in the literature today.

MUSCLE	EFFERENT INNERVATION
Sternocleidomastoid	Spinal portion of cranial XI and cervical two through four.
Scalenes	Cervical two through eight.
Pectoralis Major	Cervical five through eight. Perhaps first thoracic.
Pectoralis Minor	Cervical five through seven or eight. Perhaps first thoracic.
Subclavius	Cervical five and six.
Serratus Anterior	Cervical five through seven.
External Intercostals	Thoracic one through eleven.
Internal Intercostals	Thoracic one through eleven.
Triangularis Sterni	Thoracic two through six.
Rectus Abdominis	Thoracic seven through twelve.
External Abdominal Oblique	Thoracic eight through twelve and first lumbar.
Internal Abdominal Oblique	Thoracic eight through twelve and first lumbar.
Transverse Abdominal	Thoracic seven through twelve. Perhaps first lumbar.
Diaphragm	Cervical three through five. (Also called the phrenic nerve.)

Trapezius	Spinal portion of cranial XI and cervical two through four.
Latissimus Dorsi	Cervical six through eight.
Serratus Posterior Superior	Thoracic two and three. Perhaps first and fourth thoracic.
Serratus Posterior Inferior	Thoracic nine through twelve. Perhaps seventh and eighth thoracic.
Iliocostalis Cervicis	Branches of the cervical nerves.
Iliocostalis Thoracis	Branches of the thoracic nerves.
Iliocostalis Lumborum	Branches of the thoracic and lumbar nerves.
Costal Elevators	Cervical eight and thoracic one through eleven.
Quadratus Lumborum	Twelfth thoracic and first and second lumbar.
Subcostals	Thoracic one through eleven.

IV

THE LARYNX

The specific role of individual muscles of the larynx in voice production has not been fully determined. The larynx of man has two principle functions. One is its function in **voice production.** The other is as a sphinctor to prevent the entrance of foreign material into the lungs and to permit build up of intrathoracic pressure for such activities as coughing, vomiting, urination, and defecation. The maintenance of intrathoracic pressure is also evidently necessary to bring the full force of contraction to bear on thoracic muscles of the arms and shoulders for strenuous activity such as lifting. The sphincteric action of the larynx will hereafter be called the **biologic function** of the larynx to differentiate it from the activities of the larynx in voice production.

The larynx is essentially a tube with folds of soft tissue extending medially from its lateral internal walls. The upper rim of the larynx, on each side, is called the **aryepiglottic fold.** Some distance below this is the **ventricular fold.** More inferiorly is the **vocal fold.** The area within the larynx from the level of the aryepiglottic fold down to the level of the ventricular fold is the laryngeal **vestibule.** The area between the ventricular fold and the vocal fold is the laryngeal **ventricle.** The space between the separated vocal folds is the **glottis.** Between the ventricular and vocal folds on each side is an involution of variable size in the lateral wall extending upward toward the aryepiglottic fold. This is the laryngeal **appendix,** which may serve to lubricate the vocal folds with mucous.

The muscles of the larynx may be divided into two groups for study. Those referred to as the **extrinsic laryngeal muscles** may serve to move the larynx in the neck. The **intrinsic laryngeal muscles,** those having both attachments in or on the larynx, adjust the relative positions of the laryngeal cartilages and may effect the tension and mass of the vocal folds and the walls of the laryngeal airway. (See Fig. 65.)

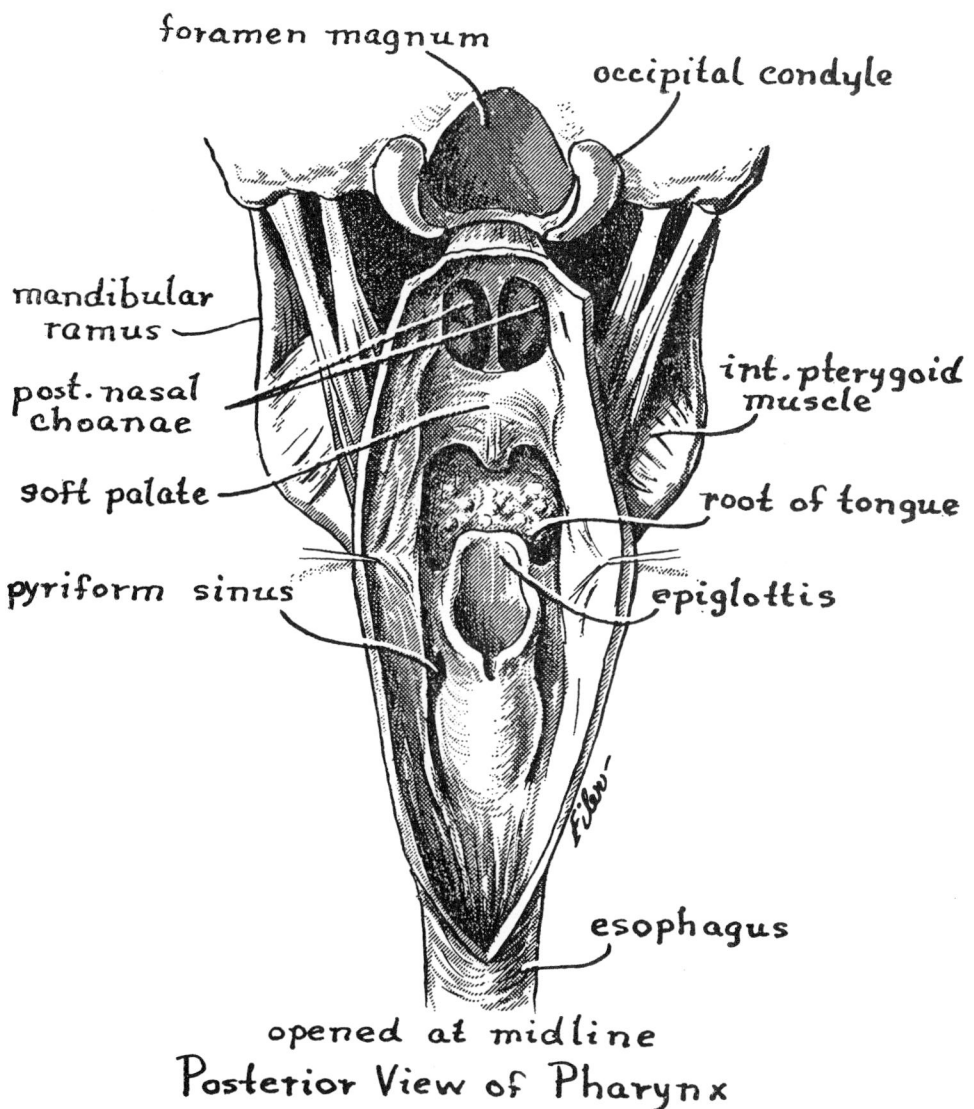

opened at midline
Posterior View of Pharynx

Figure 65.

THE LARYNGEAL CARTILAGES

CRICOID (unpaired). A ring of cartilage expanded superiorly on the sides and back, located immediately above the most superior tracheal ring. (See Fig. 66.)

LANDMARKS

Arch (unpaired). The anterior and lateral part of the cartilage.

Lamina (unpaired). The broad, flat, roughly six-sided posterior side of the cartilage.

Cricoarytenoid articular facet (unpaired). Located at the superior-lateral crest of the lamina, elongated from anterior to posterior, and doubly convex.

Cricothyroid articular facet (paired). Located at the side of the arch where it joins the lamina. Round in shape.

Posterior ridge (unpaired). Vertical ridge at midline on the posterior surface of the lamina, with a depression on either side. The ridge broadens at the base, extending across the lamina inferiorly from side to side.

ARTICULATIONS

At the superior-lateral crest of the lamina with the **arytenoid cartilages**.
Posteriorly on each side of the arch with the **thyroid cartilage**.

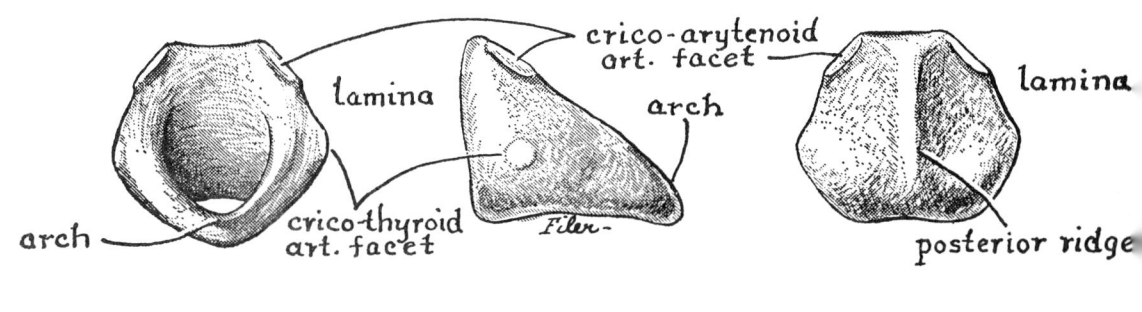

Figure 66.

THYROID (unpaired). Composed of two plates of cartilage fused anteriorly and widely separated posteriorly. (See Fig. 67.)

LANDMARKS

Lamina (paired). A roughly quadrilateral plate of cartilage lying in the lateral pharyngeal wall.

Angle (unpaired). The line of anterior fusion of the two laminae. The

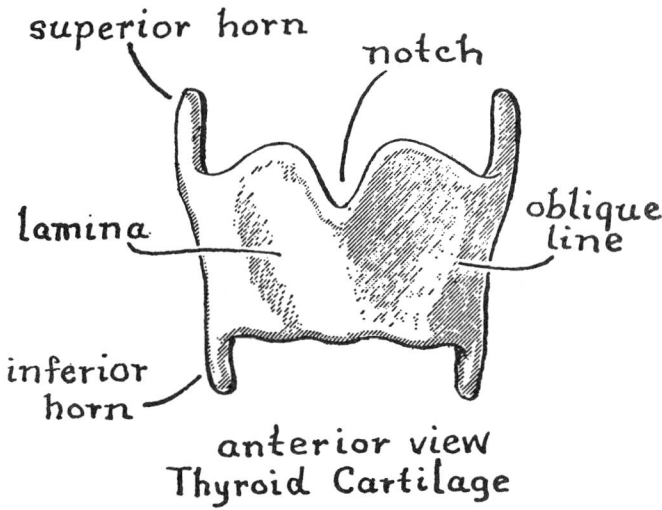

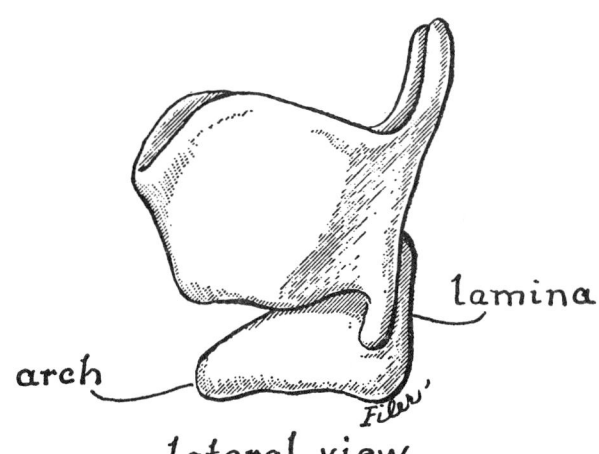

Figure 67.

anterior edges of the laminae are fused along the lower three quarters of their length.

Notch (unpaired). The separation of the two laminae immediately above the angle.

Oblique line (paired). A curved ridge on the lateral surface of the lamina, extending from its upper posterior part to its inferior border.

Superior horn (paired). Arises from the posterior limit of the superior edge of the lamina and projects toward the posterior end of the major horn of the hyoid bone.

Inferior horn (paired). Shorter and blunter than the superior horn. Descends from the posterior limit of the inferior edge of the lamina to articulate with the cricoid cartilage.

ARTICULATIONS

At the inferior horns with the **cricoid cartilage.**

EPIGLOTTIS (unpaired). Roughly oval, with the inferior end of the oval drawn to a point. Rises between the major horns of the hyoid bone, posterior to the body of the hyoid bone and the base of the tongue. The epiglottis is quite elastic and easily deformed. (See Fig. 68.)

LANDMARKS

None.

ARTICULATIONS

While the epiglottis has no true articulations, it is attached to the deep surface of the thyroid angle below the notch and to the body of the hyoid bone by the **hyoepiglottic ligament.**

ARYTENOID (paired). Complex in shape. Above its base it is roughly pyramidal with an anterolateral face, a medial face, and a dorsomedial

The Larynx 91

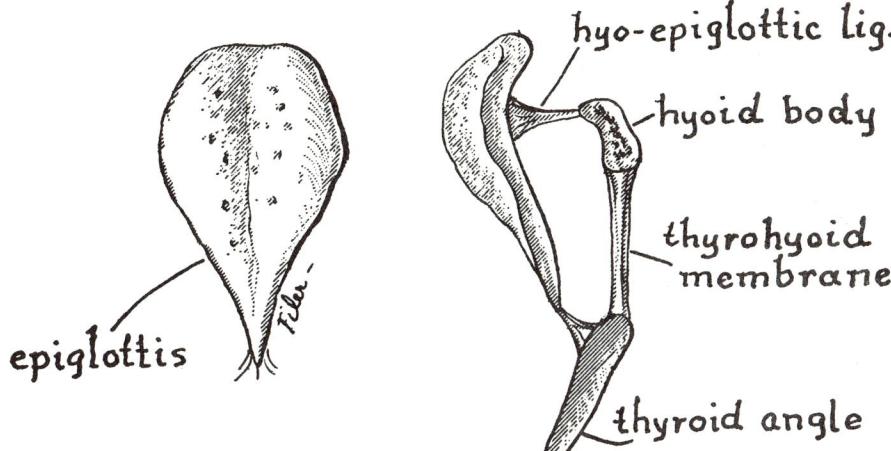

Figure 68.

face. Broad at the base, it diminishes in size as it rises toward its **apex**. Inferiorly, it is expanded into a large knoblike dorsolateral projection, the **muscular process**. Anteriorly, it has a smaller pointed anterior projection, **the vocal process**. (See Figs. 69 and 70.)

LANDMARKS

Anterolateral surface.

Medial surface.

Dorsomedial surface.

Anterior ridge. Separates the anterolateral surface from the medial surface, ending superiorly at the apex and inferiorly as the upper **edge** of the vocal process.

Posterior ridge. Separates the anterolateral surface from the dorsomedial surface, ending superiorly at the apex and inferiorly bisecting **the** muscular process.

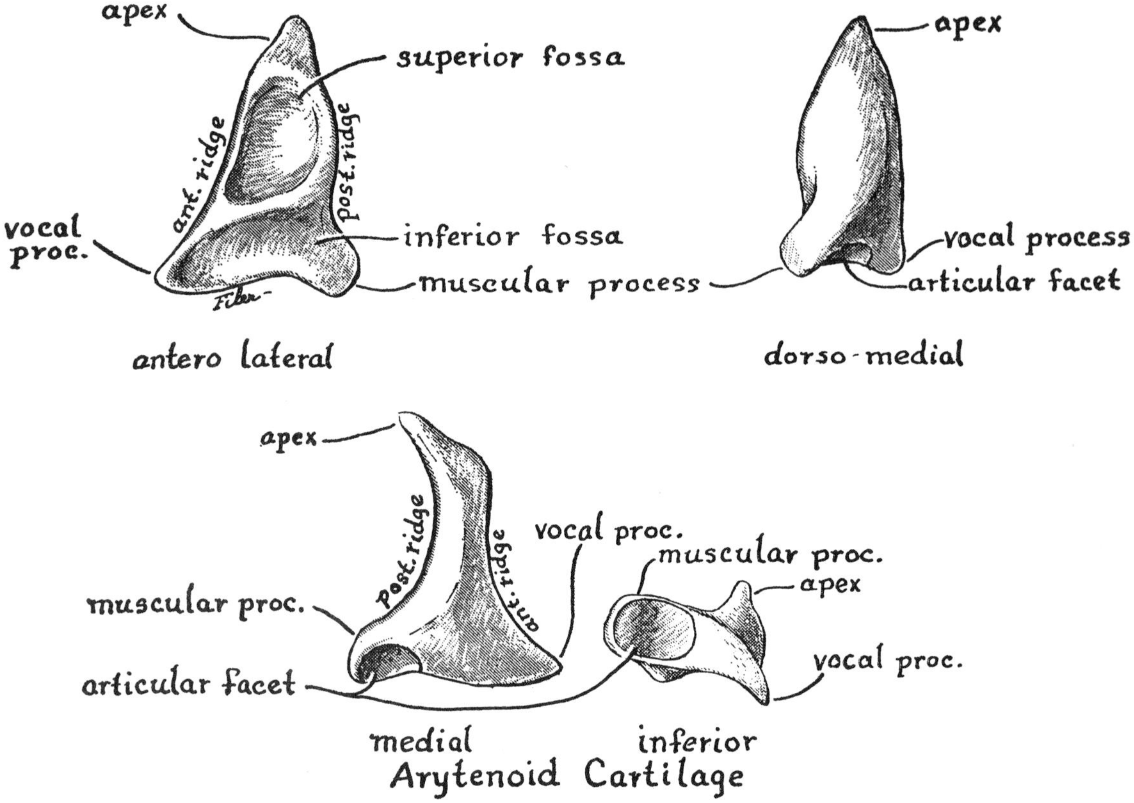

apex
superior fossa
ant. ridge
post. ridge
vocal proc.
inferior fossa
muscular process
Filer
antero lateral
apex
vocal process
articular facet
dorso-medial
apex
post. ridge
ant. ridge
vocal proc.
muscular proc.
apex
muscular proc.
vocal proc.
articular facet
medial
inferior
Arytenoid Cartilage

Figure 69.

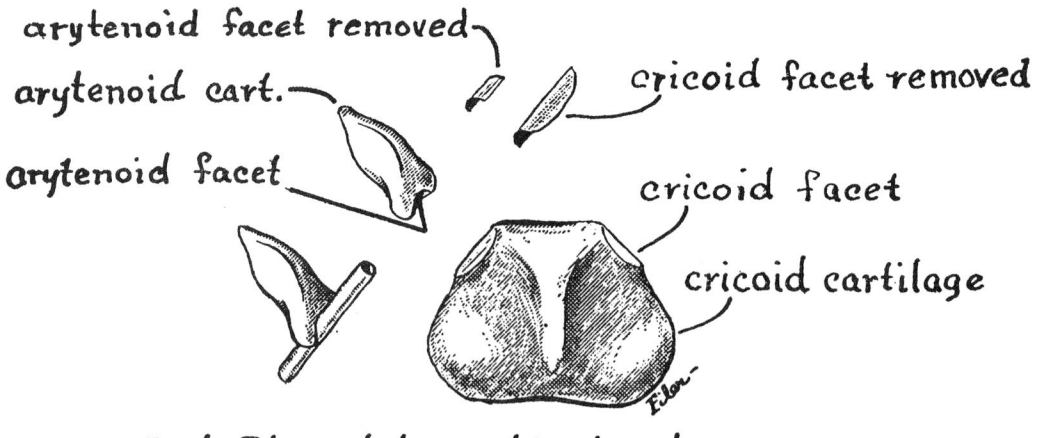

arytenoid facet removed
arytenoid cart.
cricoid facet removed
arytenoid facet
cricoid facet
cricoid cartilage
Filer
Rod Placed to indicate shape of
Articular Facet on Arytenoid Cartilage

Figure 70.

Apex. The superior extension of the arytenoid cartilage.

Vocal process. The small, pointed anterior extension from the base of the arytenoid cartilage.

Muscular process. The large knoblike dorsolateral projection of the base of the arytenoid cartilage.

Articular facet. On the underside of the muscular process, concave from dorsolateral to anteromedial. For articulation with the cricoid cartilage.

Inferior fossa. An oval depression covering the inferior half of the anterolateral surface from the vocal process to the anterior half of the muscular process.

Superior fossa. A round depression immediately above the inferior fossa.

ARTICULATIONS

At the inferior surface of the muscular process with the **cricoid cartilage.**

CORNICULATE (paired). A small elastic cartilage superior to and sometimes fused with the apex of the arytenoid. Extends the apex dorsomedially.

CUNEIFORM (paired). A small, rod-shaped cartilage. Variable in number. Lies in the aryepiglottic fold just lateral to the corniculate cartilage.

LIGAMENTS OF THE LARYNX

The **ligaments** of the larynx and their expanded form, the **membranes,** serve to attach the several cartilages to one another to provide part of the supporting structure of the larynx.

THYROHYOID MEMBRANE (unpaired). Extends from the entire superior margin of the thyroid cartilage, including the superior horns, to the entire inferior surface of the body and major horns of the hyoid bone. (See Fig. 71.)

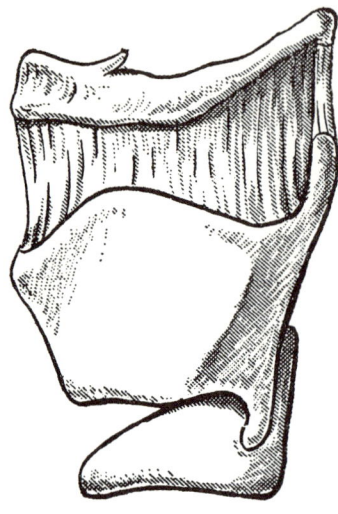 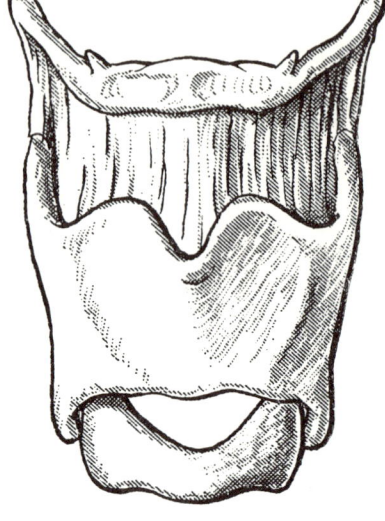

Lateral view   Anterior view
Thyrohyoid Membrane

Figure 71.

CRICOTHYROID MEMBRANE (paired). Complex in form and composed of three parts. (See Fig. 72.)

Anterior cricothyroid ligament (unpaired).

Extends vertically at midline from the superior margin of the anterior-most part of the cricoid arch to the inferior margin and deep surface of the thyroid angle.

Vocal ligament (paired).

Extends from the vocal process of the arytenoid cartilage to the deep surface of the thyroid angle.

Conus elasticus (paired).

Attaches inferiorly to the length of the superior border of the cricoid arch and curves superiorly and medially to end in an upper thickened border, the vocal ligament. Ends anteriorly in the cricothyroid ligament.

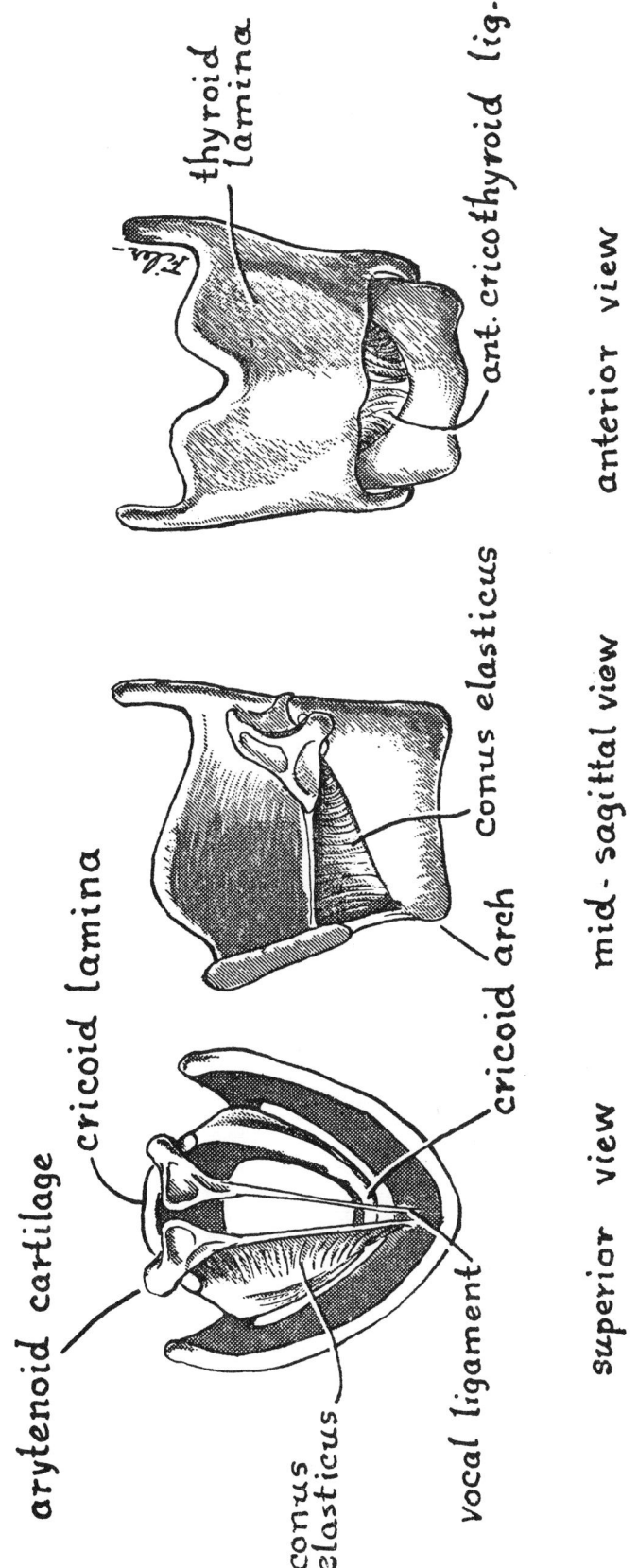

Figure 72.

Cricothyroid Membrane

QUADRANGULAR MEMBRANE (paired). Located in the lateral wall of the laryngeal vestibule. Roughly four sided. The superior edge lies in the **ary**epiglottic fold. The inferior edge is thickened to form the **ventricular ligament,** which lies in the ventricular fold and extends from the superior fossa of the arytenoid cartilage to the deep surface of the thyroid angle. The anterior edge attaches to the entire lateral edge of the epiglottis. The posterior edge attaches to the superior part of the arytenoid cartilage. The deep surface of the membrane faces the laryngeal airway, and the superficial surface faces the **pyriform sinus** (the lateral expansions of the laryngeal pharynx). (See Fig. 73.)

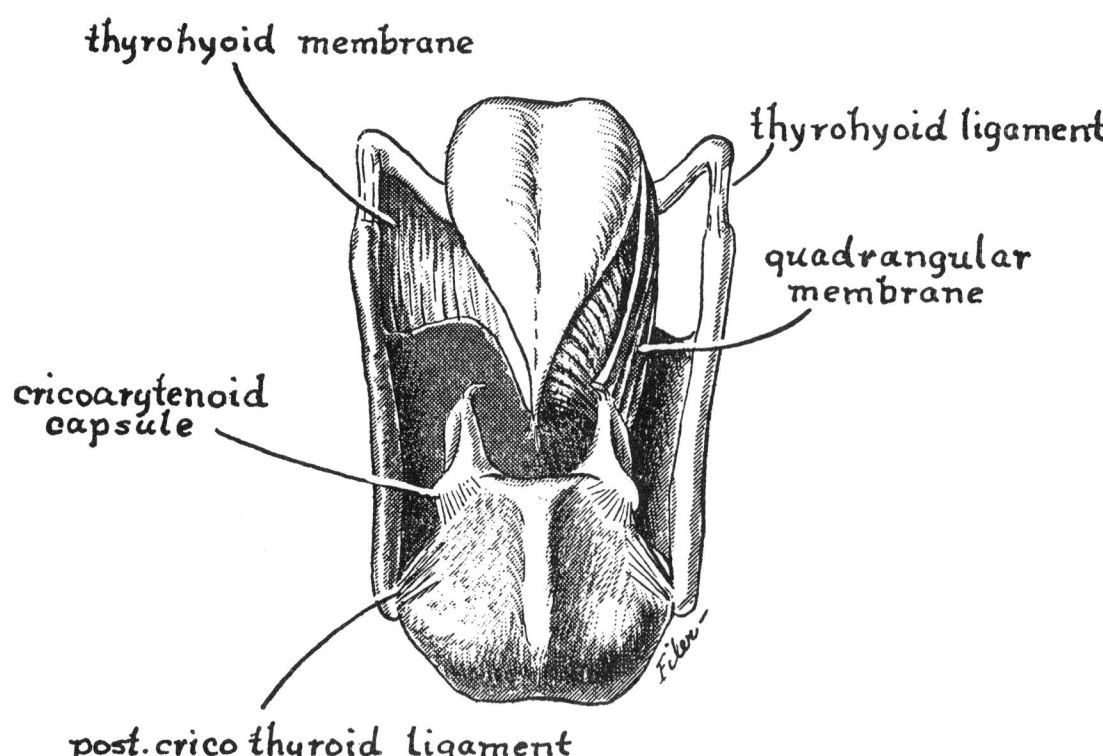

posterior view
Laryngeal Ligaments

Figure 73.

THYROHYOID LIGAMENT (paired). The thickened portion of the thyrohyoid membrane between the superior horn of the thyroid cartilage and the posterior end of the major horn of the hyoid bone.

HYOEPIGLOTTIC LIGAMENT (unpaired). Attaches to the anterior surface of the epiglottis and to the posterior surface of the body of the hyoid bone.

POSTERIOR CRICOTHYROID LIGAMENT. Not well understood. Form a capsule which encloses the cricothyroid articulation. (See Fig. 74.)

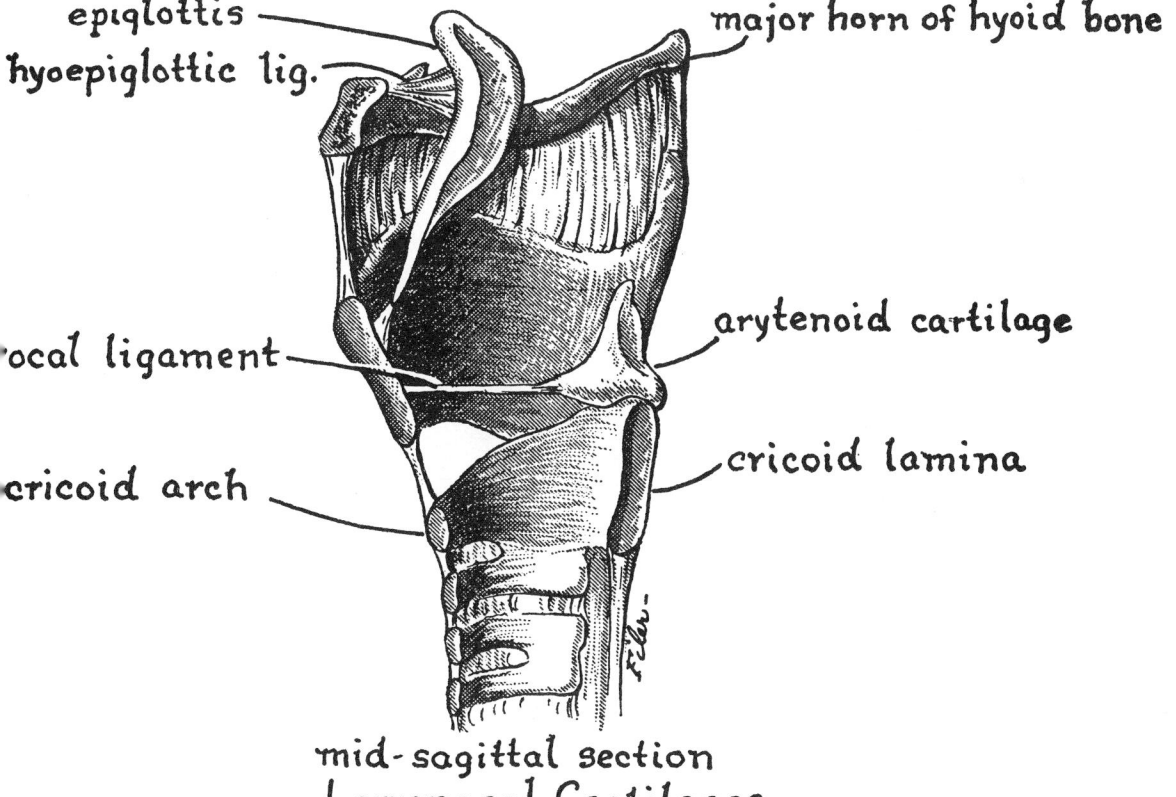

Figure 74.

CRICOARYTENOID LIGAMENT. Not well understood. Form a capsule which encloses the cricoarytenoid articulation. A band of ligament, the **posterior cricoarytenoid ligament,** extends from the dorsomedial surface of the arytenoid cartilage to the posterior surface of the cricoid lamina. (See Fig. 75.)

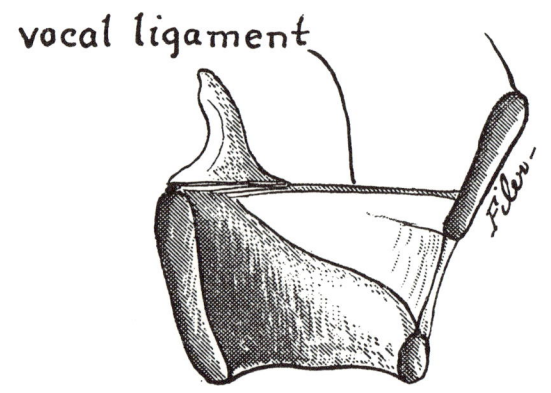

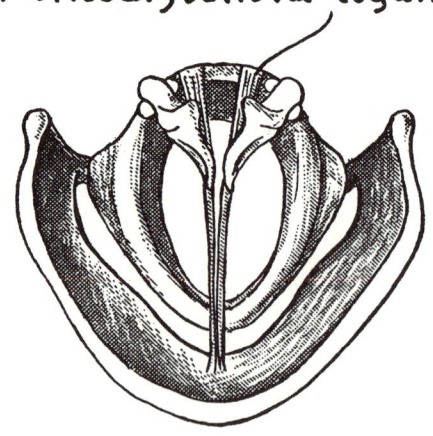

Figure 75.

EXTRINSIC MUSCLES OF THE LARYNX
Suprahyoid Muscles

DIGASTRIC (paired). A well-developed muscle consisting of an anterior belly and a posterior belly connected by an intermediate tendon. The anterior belly is the most inferior muscle of the floor of the mouth. The posterior belly is deep to the sternocleidomastoid muscle. (See Fig. 76.)

ATTACHMENTS

To the deep surface of the body of the mandible just lateral to the mandibular symphysis.

By a loop of ligament which passes over the intermediate tendon, to the hyoid body near its minor horn.

To the mastoid process of the temporal bone deep to the sternocleidomastoid muscle.

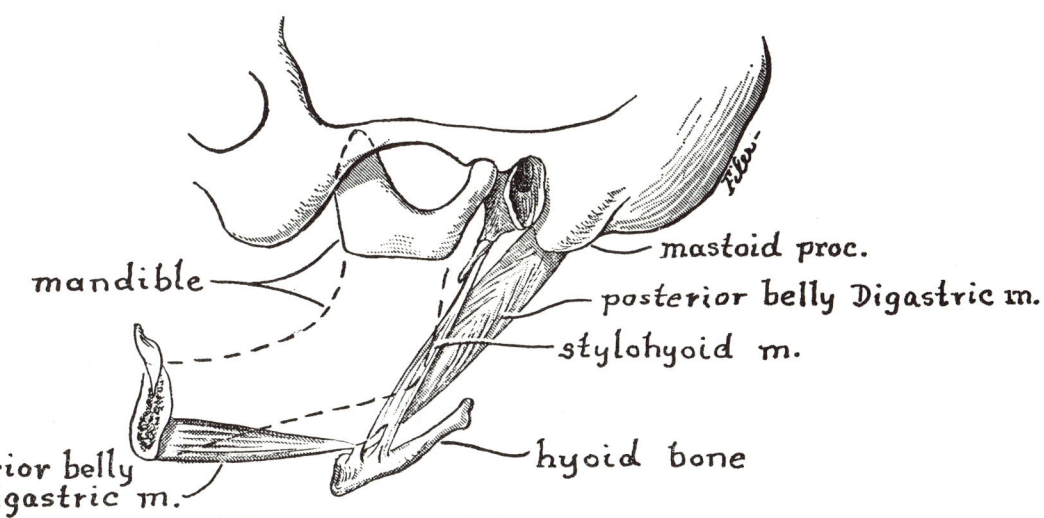

Lateral view, mandible cut
Digastric and Stylohyoid Muscles

Figure 76.

FUNCTION

The anterior belly aids in lowering the mandible and, acting alone, will draw the hyoid bone forward and slightly upward.

The posterior belly alone will draw the hyoid bone upward and posteriorly.

Together the bellies will raise the hyoid bone and indirectly, the larynx.

MYLOHYOID (paired). A flat sheet of muscle which lies superior to the anterior belly of the digastric muscle. The two sides together form the muscular floor of the oral cavity. The fibers of the two sides extend from the body of the mandible downward and toward the midline. (See Fig. 77.)

ATTACHMENTS

To the mylohyoid line on the deep surface of the body of the mandible.

To the median raphe, which extends from the mandibular symphysis posteriorly to the center of the body of the hyoid bone.

To the anterior surface of the body of the hyoid bone.

FUNCTION

May aid in lowering the mandible.

May draw the hyoid bone forward and upward.

GENIOHYOID (paired). Lies superior to the mylohyoid muscle. Composed of well-developed trunks of muscle on either side of midline which extend from the mandibular symphysis to the hyoid bone. (See Fig. 78.)

ATTACHMENTS

To the inferior mental spine of the mandible.

To the anterior surface of the body of the hyoid bone adjacent to midline.

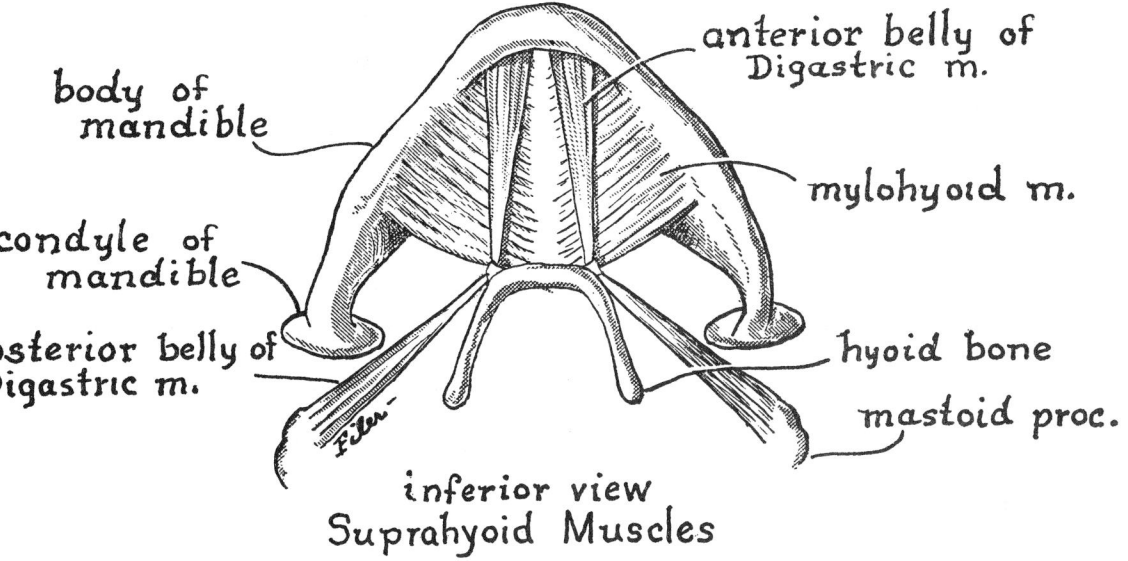

Figure 77.

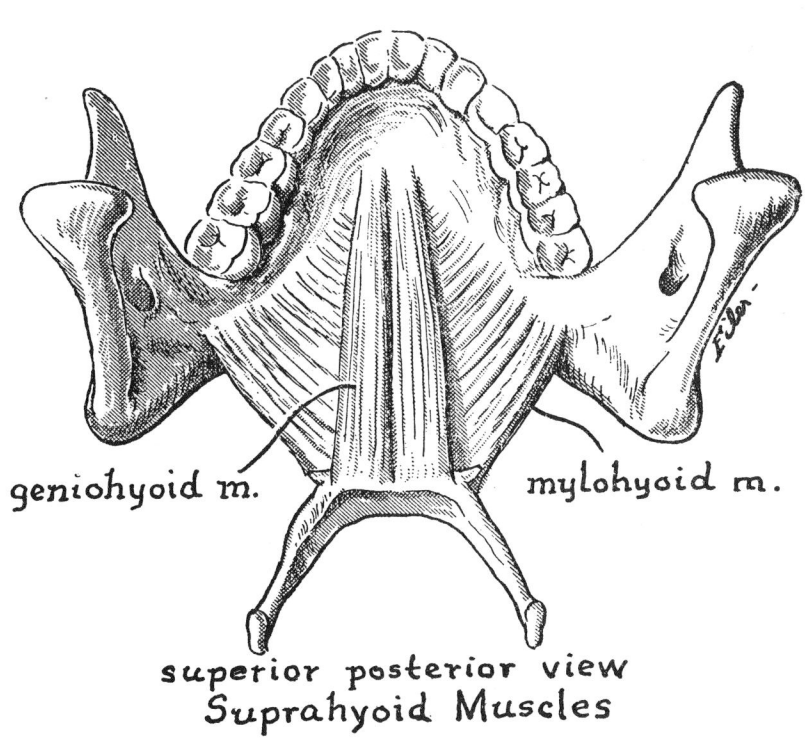

Figure 78.

FUNCTION

May aid in lowering the mandible.

Draws the hyoid bone forward and perhaps slightly upward.

STYLOHYOID (paired). A slender muscle roughly parallel to the posterior belly of the digastric muscle. (See Fig. 76.)

ATTACHMENTS

To the styloid process of the temporal bone by a long tendon.

Divides, one half passing on either side of the intermediate tendon of the digastric muscle to attach to the body of the hyoid bone near its minor horn.

FUNCTION

Draws the hyoid bone superiorly and posteriorly.

Infrahyoid Muscles

STERNOHYOID (paired). A thin, flat muscle lying superficially in the anterior portion of the neck on either side of midline. (See Fig. 79.)

ATTACHMENTS

To the posterior-superior part of the manubrium of the sternum and to the medial end of the clavicle.

To the lower border of the body of the hyoid bone near midline.

FUNCTION

Draws the hyoid bone inferiorly.

OMOHYOID (paired). A thin, flat muscle lateral to the sternohyoid muscle. Consists of a superior belly and an inferior belly joined by a central tendon.

ATTACHMENTS

To the superior border of the scapula lateral to its medial angle.

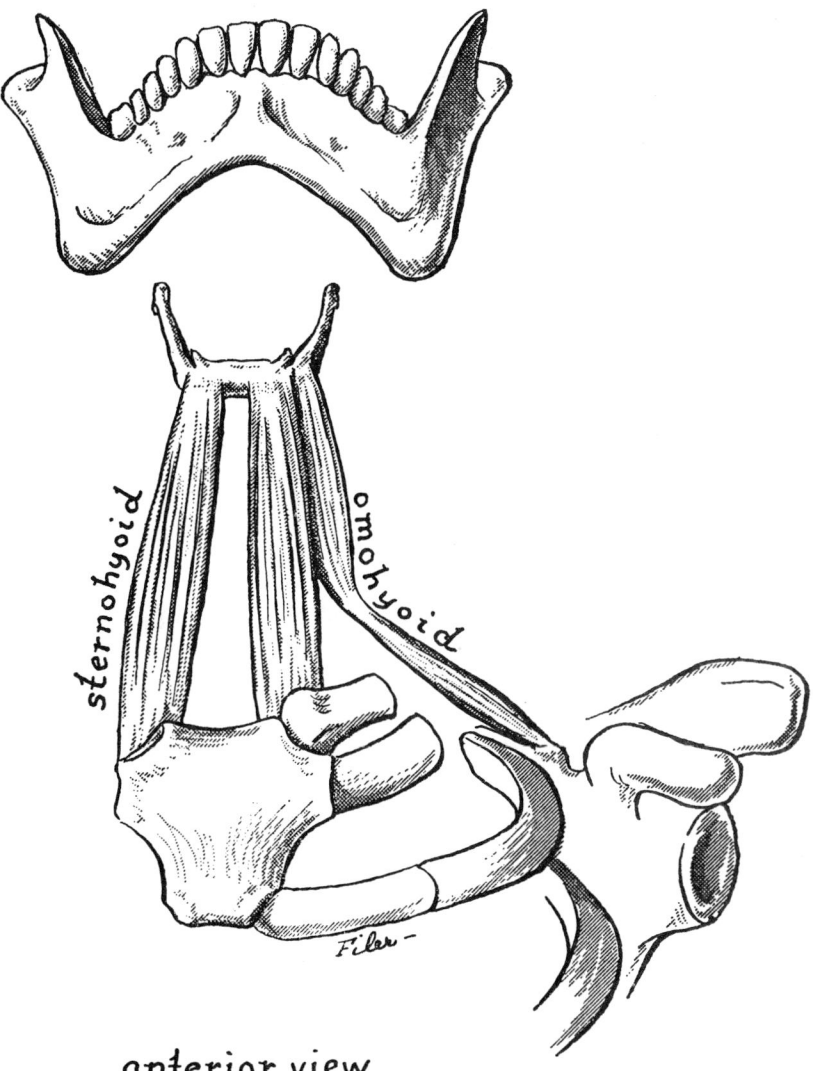

anterior view
Sternohyoid and Omohyoid Muscles

Figure 79.

Extends forward and upward to a point above the medial end of the clavicle. It is held in position by fascia extending upward from the clavicle.

To the inferior border of the body of the hyoid bone lateral to the sternohyoid muscle.

FUNCTION

Draws the hyoid bone inferiorly and perhaps somewhat posteriorly.

STERNOTHYROID (paired). A flat muscle deep to the sternohyoid muscle. (See Fig. 80.)

ATTACHMENTS

To the deep surface of the manubrium of the sternum and the medial end of the cartilage of the first rib.

To the oblique line of the thyroid lamina.

FUNCTION

Draws the thyroid cartilage downward.

THYROHYOID (paired). A flat muscle deep to the upper portion of the sternohyoid muscle.

ATTACHMENTS

To the oblique line of the thyroid lamina.

To the lower border of the body and greater horn of the hyoid bone.

FUNCTION

Approximates the hyoid bone and the thyroid cartilage.

INTRINSIC MUSCLES OF THE LARYNX

CRICOTHYROID (paired). Roughly triangular in form and well developed. Lies on the superficial surface of the side of the larynx. (See Fig. 81.)

ATTACHMENTS

To the anterior and lateral surface of the cricoid arch.

The anterior or **erect portion** rises to attach to the inferior margin of the thyroid cartilage.

The posterior or **oblique portion** is directed superiorly and posteriorly to attach to the superficial and deep surfaces of the inferior horn of the thyroid cartilage.

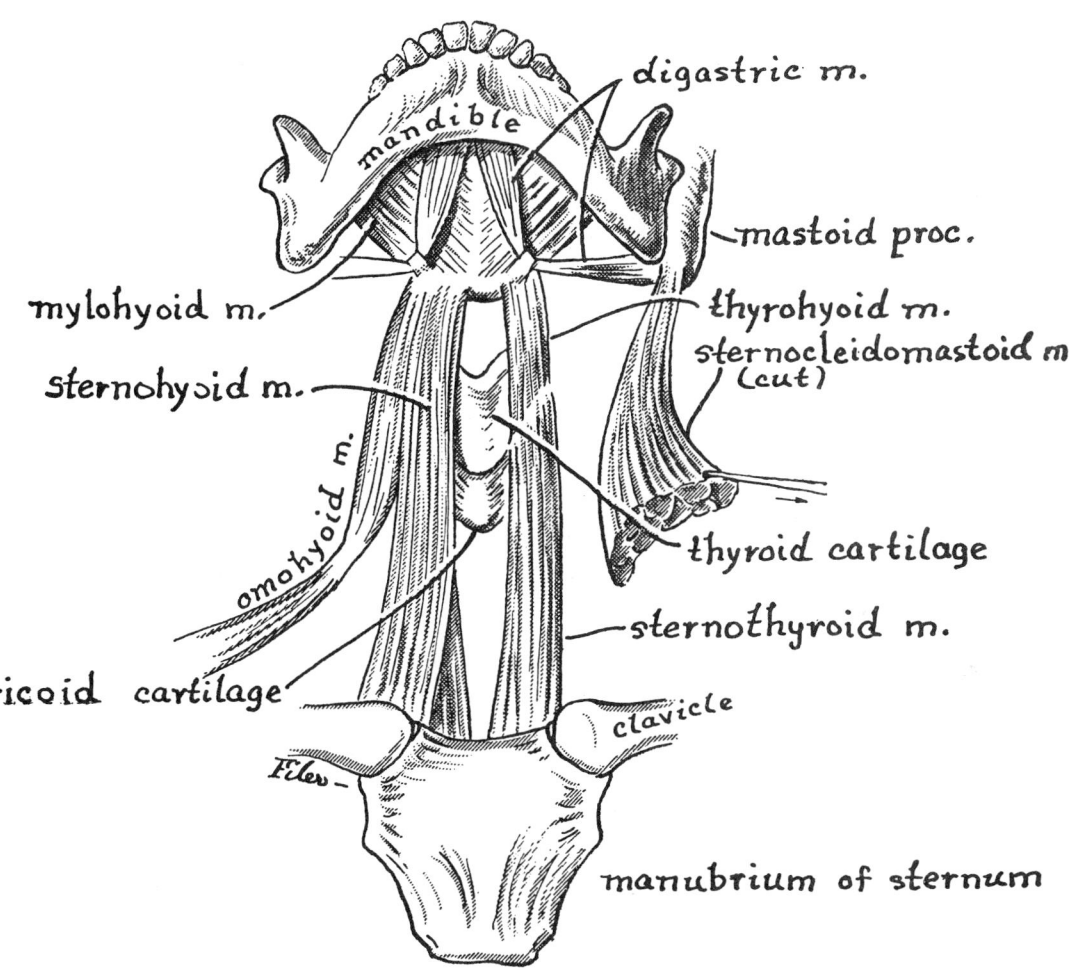

anterior view
Extrinsic Muscles of the Larynx

Figure 80.

FUNCTION

Lifts the anterior arch of the cricoid cartilage, tilting the top of the lamina posteriorly, thus lengthening the vocal folds. May also move the cricoid cartilage posteriorly in relation to the thyroid cartilage.

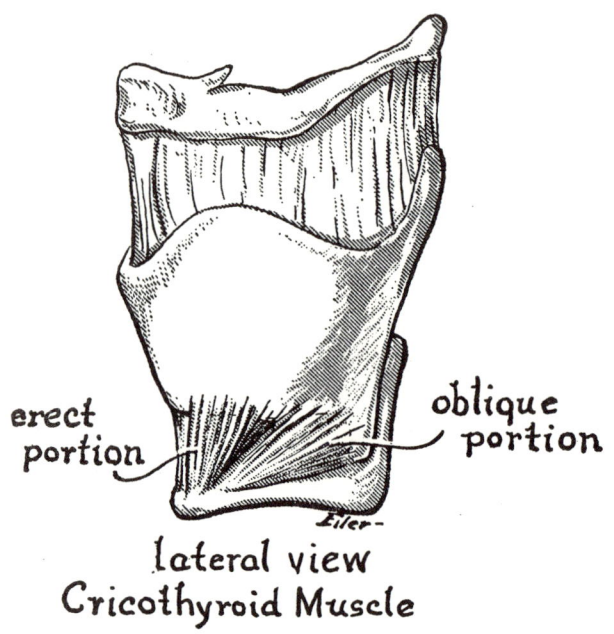

Figure 81.

POSTERIOR CRICOARYTENOID (paired). A well-developed triangular muscle on the dorsal surface of the cricoid lamina. (See Fig. 82.)

ATTACHMENTS

To the curved depression on the dorsal surface of the cricoid lamina lateral to the posterior ridge.

Fibers converge to attach by a broad thin tendon to the superior surface of the muscular process of the arytenoid cartilage.

FUNCTION

Moves the arytenoid cartilage in a rocking movement posterolaterally over the rim of the cricoid cartilage, thus abducting the vocal folds.

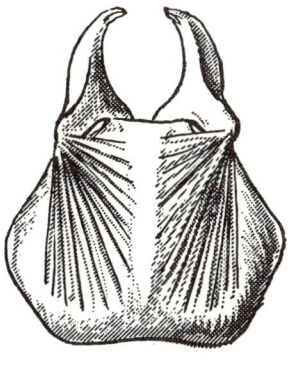

posterior view

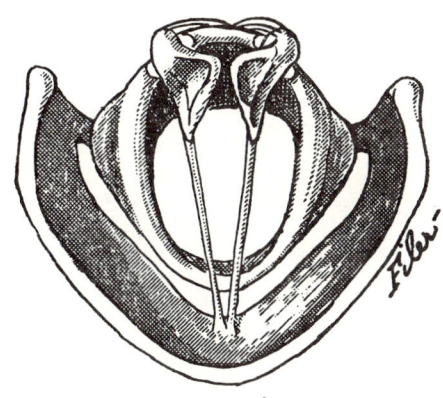

superior view

Posterior Cricoarytenoid Muscle

Figure 82.

TRANSVERSE ARYTENOID (unpaired). Its fibers extend horizontally, covering the posterior surfaces of the arytenoid cartilages and filling the concavity between the muscular processes and apexes. (See Fig. 83.)

ATTACHMENTS

To the entire length of the posterior ridge of each arytenoid cartilage from the apex to the muscular process. Some fibers are continuous with fibers of the lateral thyroarytenoid muscle.

FUNCTION

Aids in adduction of the arytenoid cartilages.

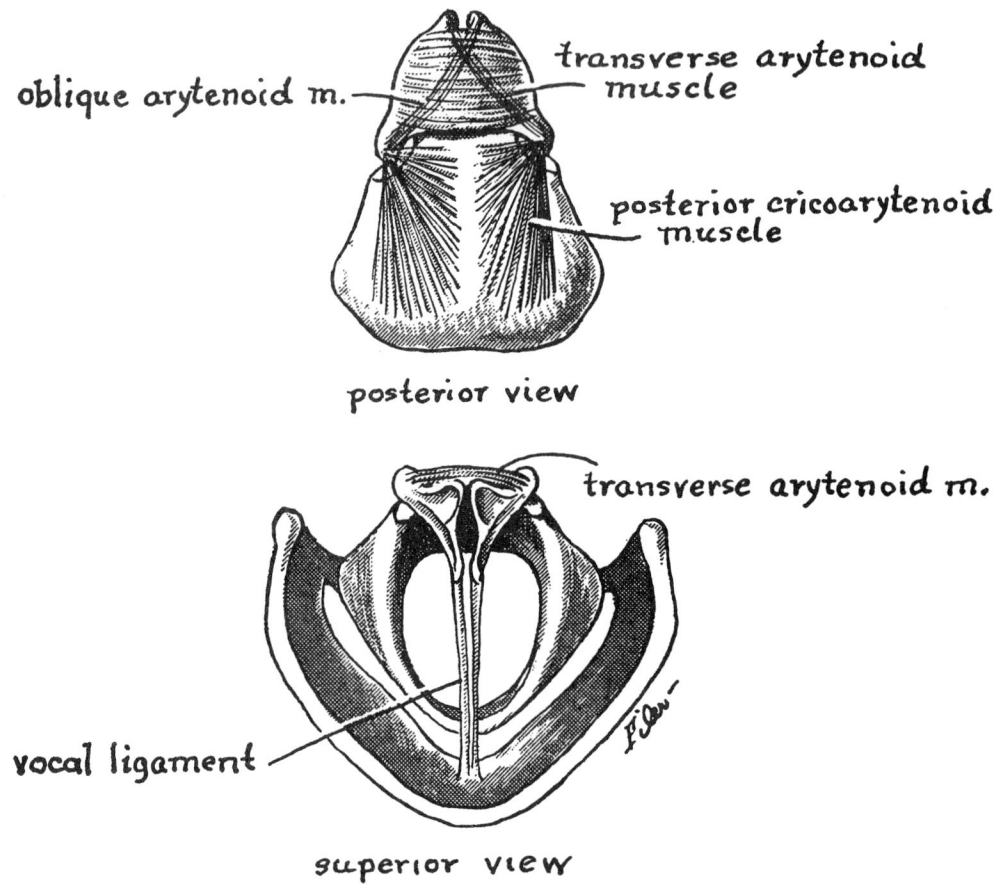

Posterior Laryngeal Muscles

Figure 83.

OBLIQUE ARYTENOID (paired). Its fibers form an "X" across the posterior surfaces of the arytenoid cartilages dorsal to the transverse arytenoid muscle.

ATTACHMENTS

To the apex of one arytenoid cartilage and to the medial half of muscular process of the opposite arytenoid cartilage. Frequently its fibers are continuous with fibers of the aryepiglottic muscle.

FUNCTION

May aid in adduction of the arytenoid cartilages. May aid in closure of the superior aperture of the larynx.

LATERAL CRICOARYTENOID (paired). Triangular in form. Lies deep to the thyroid cartilage. Its fibers parallel adjacent fibers of the lateral thyroarytenoid muscle. (See Fig. 84.)

ATTACHMENTS

To the lateral portion of the superior rim of the arch of the cricoid cartilage. Some fibers may attach to the conus elasticus.

To the anterior half of the muscular process of the arytenoid cartilage.

FUNCTION

May draw the arytenoid cartilage anteriorly, shortening the vocal folds when they are held in adduction.

May aid in adduction of the arytenoid cartilages.

May cause slight rotation of the arytenoid cartilages, pressing the two vocal processes together.

VOCALIS (paired). Sometimes called the medial portion of the thyroarytenoid muscle. Well developed. Lies lateral to the vocal ligament. (See Fig. 85.)

ATTACHMENTS

To the anterior half of the inferior fossa and adjacent portion of the vocal process of the arytenoid cartilage.

To the deep surface of the lamina of the thyroid cartilage adjacent to the attachment of the vocal ligament.

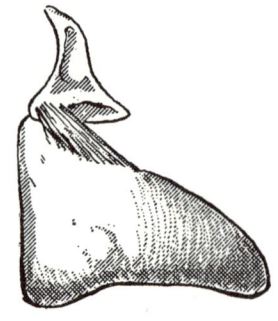

lateral view

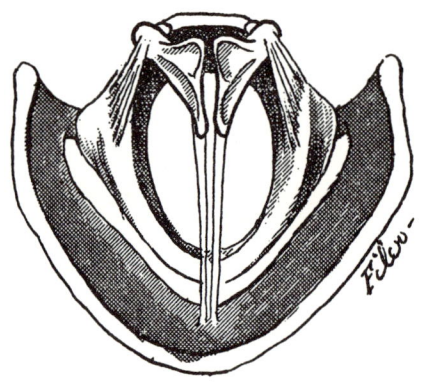

superior view

Lateral Cricoarytenoid Muscle

Figure 84.

FUNCTION

For biologic function, may aid in vocal fold adduction and shortening. For phonation, increases tension in the vocal folds.

THYROARYTENOID (paired). A well-developed muscle lateral to the vocalis muscle. Extends upward into the lateral wall of the ventricle. (See Figs. 86 and 87.)

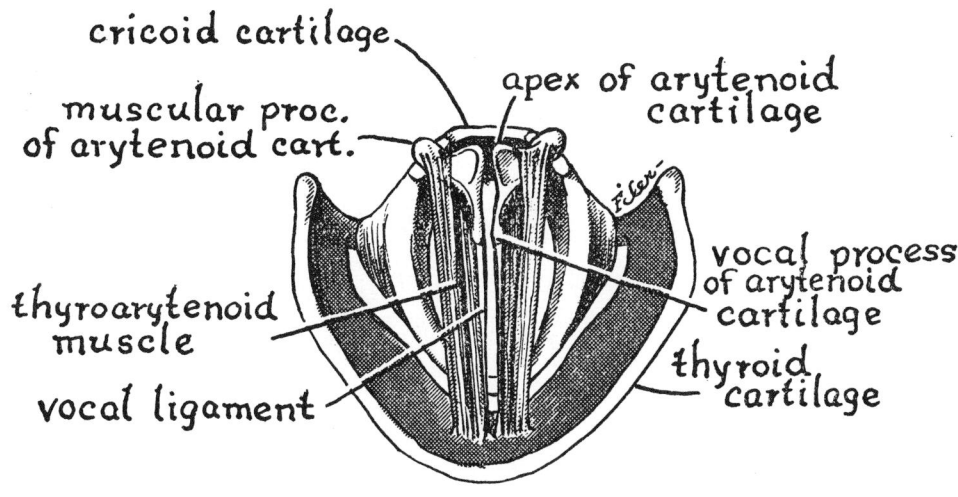

superior view
Thyroarytenoid Muscle

Figure 85.

ATTACHMENTS

Its medial portion attaches to the lateral half of the inferior fossa and the anterior half of the muscular process of the arytenoid cartilage. Its lateral portion attaches to the dorsal ridge of the arytenoid cartilage and extends up toward the apex, with some fibers continuous with the transverse arytenoid muscle.

To the deep surface of the lamina of the thyroid cartilage lateral to the attachment of the vocalis muscle.

FUNCTION

The lateral portion may act with the transverse arytenoid muscle in adduction of the arytenoid cartilages. May shorten the adducted vocal folds during closure for biologic or phonatory purposes.

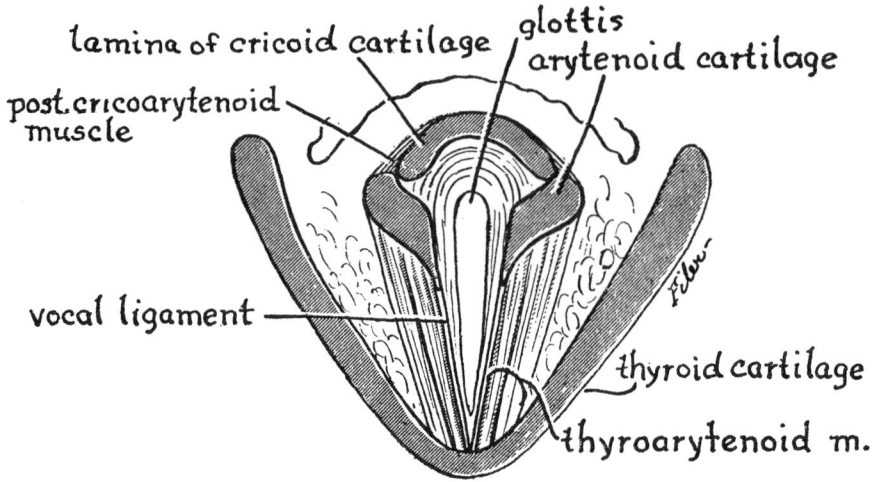

Transverse Section of Larynx

Figure 86.

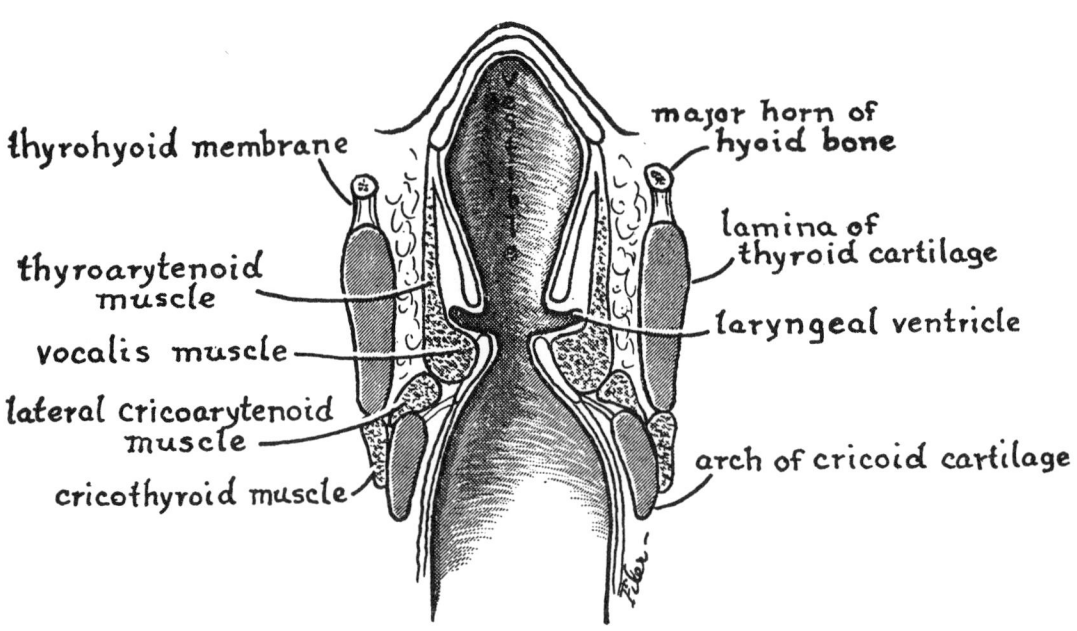

Frontal Section of the Larynx

Figure 87.

The medial portion may aid in adduction of the arytenoid cartilages and shortening of the vocal folds for biologic purposes. May aid in tension regulation for phonation.

The muscle acting as a whole may aid in vocal fold adduction, shortening, or tension regulation.

VENTRICULAR (paired). An inconstant muscle. Not well developed. Its fibers course through the ventricular fold.

ATTACHMENTS

Not clear but appear to be to the superior fossa of the arytenoid cartilage and to the deep surface of the lamina of the thyroid cartilage superior to the attachment of the vocalis muscle.

FUNCTION

May aid in sphincteric closure for biologic purposes.

ARYEPIGLOTTIC (paired). An inconstant muscle. Not well developed. Its fibers course through the aryepiglottic fold. Some of its fibers seem to be continuations of fibers of the oblique arytenoid muscle. (See Fig. 88.)

ATTACHMENTS

Not clear but appear to be to the apex of the arytenoid cartilage and to the lateral margin of the epiglottis superiorly.

FUNCTION

May aid in sphincteric closure for biologic purposes.

THYROEPIGLOTTIC (paired). Lies superior to the lateral portion of the thyroarytenoid muscle. May be an upward extension of the thyroarytenoid muscle.

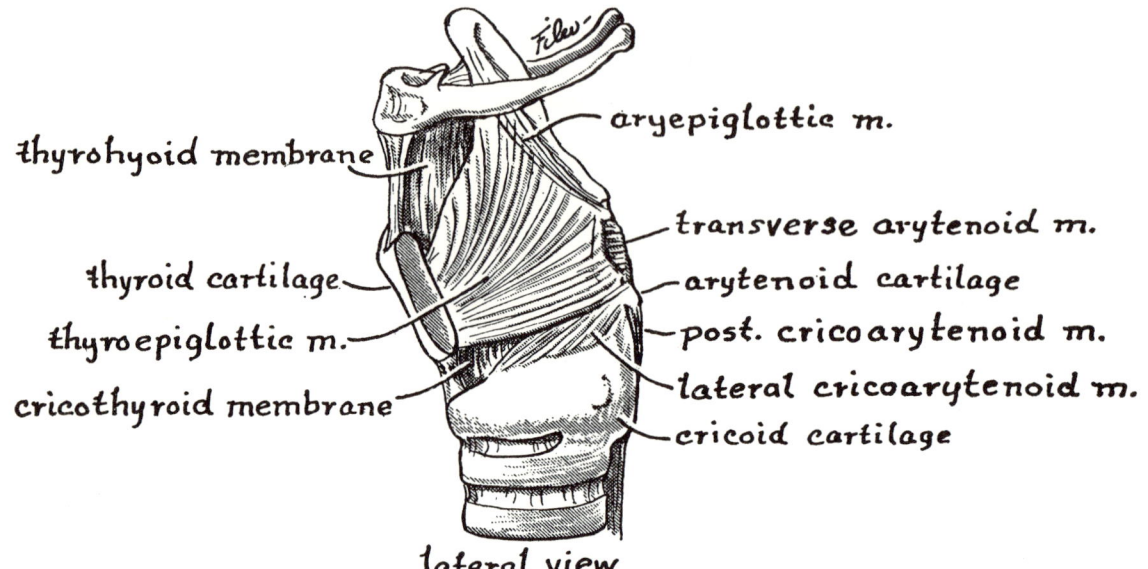

Larynx with the Thyroid Lamina Removed

Figure 88.

ATTACHMENTS

To the deep surface of the lamina of the thyroid cartilage superior to the attachment of the lateral portion of the thyroarytenoid muscle.

To the posterior half of the aryepiglottic fold.

FUNCTION

May aid in sphincteric closure for biologic purposes.

PERIPHERAL INNERVATION OF THE MUSCLES OF THE LARYNX

Branches of cranial nerves V, VII, and XII supply the extrinsic laryngeal muscles. The intrinsic muscles of the larynx are supplied by branches of cranial nerve X. The exact distribution of these nerves is not clear. The following list of efferent innervation represents what seems to be the consensus in the literature today.

The Larynx

MUSCLE	EFFERENT INNERVATION
Digastric	
Anterior Belly:	Mandibular Branch of Cranial V.
Posterior Belly:	Cervicofacial Branch of Cranial VII.
Mylohyoid	Mandibular Branch of Cranial V.
Geniohyoid	Cranial XII.
Stylohyoid	Cervicofacial Branch of Cranial VII.
Sternohyoid	Cranial XII, perhaps upper cervical nerves.
Omohyoid	Cranial XII.
Sternothyroid	Cranial XII, perhaps upper cervical nerves.
Thyrohyoid	Cranial XII, perhaps upper cervical nerves.
Cricothyroid	Superior Laryngeal Nerve from Cranial XI.
Posterior Cricoarytenoid	Recurrent Laryngeal Nerve from Cranial XI.
Transverse Arytenoid	Recurrent Laryngeal Nerve from Cranial XI.
Oblique Arytenoid	Recurrent Laryngeal Nerve from Cranial XI.
Lateral Cricoarytenoid	Recurrent Laryngeal Nerve from Cranial XI.
Vocalis	Recurrent Laryngeal Nerve from Cranial XI.
Thyroarytenoid	Recurrent Laryngeal Nerve from Cranial XI.
Ventricular	Recurrent Laryngeal Nerve from Cranial XI.
Aryepiglottic	Recurrent Laryngeal Nerve from Cranial XI.
Thyroepiglottic	Recurrent Laryngeal Nerve from Cranial XI.

V

ARTICULATION

Articulation of sounds involves the soft palate (velum) and pharynx in velopharyngeal closure and movements of the tongue, mandible, and lips. All of the musculature of these parts is also involved in the acts of chewing and swallowing. While the musculature involved in mandibular movements and lip movements is fairly well understood, there is a good deal of debate concerning the muscles responsible for velopharyngeal closure and specific movements of the tongue.

THE SOFT PALATE AND PHARYNX

The pharynx is a muscular tube incomplete in the front where it is continuous with the openings into the nasal, oral, and laryngeal cavities. It extends from the base of the sphenoid bone above to the entrance into the esophagus at the level of the cricoid cartilage below. It is wider above than below. The part of the pharynx behind the nasal cavities from the body of the sphenoid bone to the soft palate is the **nasopharynx.** The **oropharynx,** located behind the oral cavity, extends from the soft palate above to the level of the hyoid bone below. The **laryngopharynx,** from the hyoid bone to the esophagus, is broad but has a small anteroposterior dimension behind the larynx and a large out-pouching, the **pyriform sinus,** on either side of the larynx.

The soft palate is composed of muscle, connective, and gland tissues. It is attached to the posterior rim of the hard palate anteriorly, to the side walls of the pharynx and oral cavity laterally, and hangs freely into the pharynx behind the tongue posteriorly.

Velopharyngeal closure for speech apparently involves the upward and backward movement of the soft palate against the posterior pharyngeal wall about the level of the anterior tubercle of the atlas. Medial movement of the lateral pharyngeal walls against the sides of the soft palate completes velopharyngeal closure. Occasionally, especially in persons with short or cleft palates, a ledge of tissue is seen to bulge forward from the posterior pharyngeal wall toward the soft palate. This bulge is termed **Passavant's Pad** and is thought to be a compensatory mechanism.

MUSCLES OF THE PHARYNX

SUPERIOR CONSTRICTOR (paired). Its fibers extend from midline of the posterior pharyngeal wall, from the level of the sphenoid body to the level of the body of the mandible, downward and forward to a complex series of anterior attachments. (See Figs. 89 and 90.)

ATTACHMENTS

To the median pharyngeal raphe. The most superior fibers arch downward and forward, leaving a superior space laterally through which the levator palatini muscle and the auditory tube pass.

To the lower third of the medial pterygoid plate, hamulus, pterygomandibular raphe (a line of connective tissue from the hamulus to the posterior end of the mylohyoid line), and to the mandible just superior to mylohyoid line.

Some fibers of the superior constrictor muscle to enter the tongue. Other fibers of this muscle, at least in some cases, enter the soft palate.

FUNCTION

Constricts the area of the pharynx in swallowing.

May aid in movement of the lateral walls of the pharynx in speech.

May aid in posterior movements of the tongue.

MIDDLE CONSTRICTOR (paired). Somewhat thicker than the superior constrictor muscle. Located superficial (lateral and posterior) to the superior constrictor muscle.

ATTACHMENTS

To the median pharyngeal raphe. The most superior fibers attach somewhat lower than the uppermost fibers of the superior constrictor

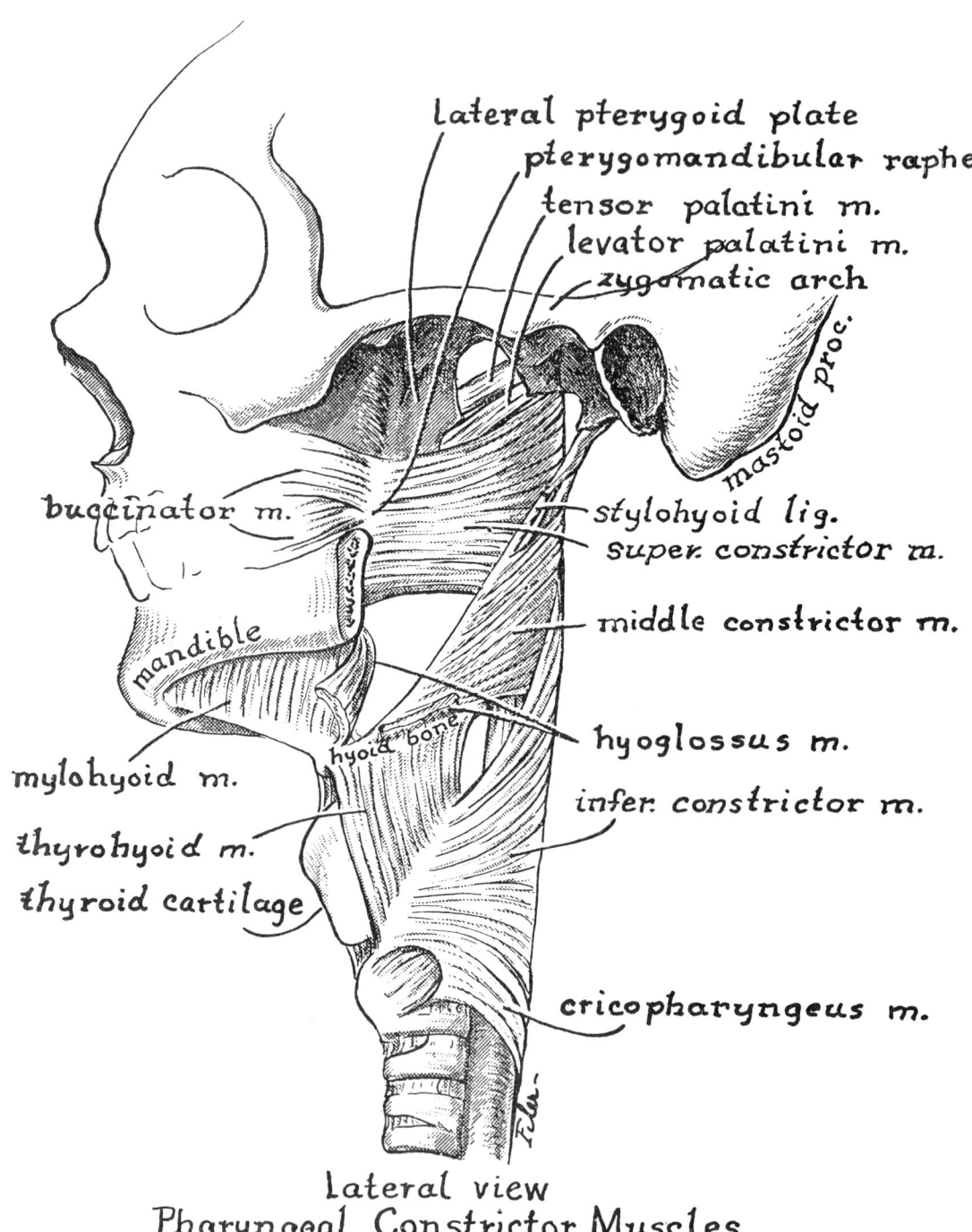

Lateral view
Pharyngeal Constrictor Muscles

Figure 89.

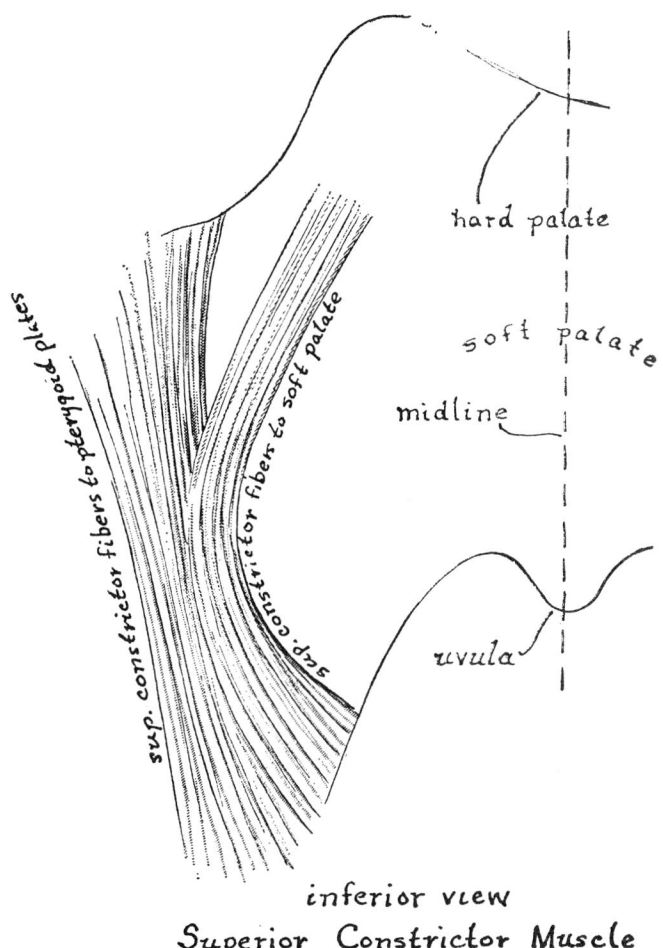

inferior view
Superior Constrictor Muscle

Figure 90.

muscle. Its fibers are directed more inferiorly, with the most inferior fibers at or below the level of the hyoid bone.

To the superior border of the greater horns of the hyoid bone, the lesser horns of the hyoid bone, and the stylohyoid ligament.

FUNCTION

Constricts the area of the pharynx during swallowing.

May move the hyoid bone posteriorly.

INFERIOR CONSTRICTOR (paired). The thickest of the three constrictor muscles. Located superficial (lateral and posterior) to the middle constrictor muscle. The most inferior fibers, which attach to the cricoid cartilage, are horizontal and are often called the **cricopharyngeus muscle.**

ATTACHMENTS

To the median pharyngeal raphe. The most superior fibers attach somewhat lower than the uppermost fibers of the middle constrictor muscle. Its fibers are directed more inferiorly, with the most inferior fibers at the level of the entrance into the esophagus.

To the lateral margins of the cricoid and thyroid cartilages.

FUNCTION

Constricts the area of the pharynx during swallowing.

May move the larynx up and back.

The cricopharyngeus part is felt to be the sphincter used for esophageal speech.

STYLOPHARYNGEUS (paired). A muscle cylindrical superiorly and flattened inferiorly in the lower pharyngeal wall. Passes between the superior and middle constrictor muscles into the lateral pharyngeal wall. (See Fig. 91.)

ATTACHMENTS

To the styloid process of the temporal bone.

Some fibers may reach the lamina of the thyroid cartilage inferiorly.

FUNCTION

Raises and widens the pharynx.

SALIPINGOPHARYNGEUS. See Muscles of the Soft Palate.

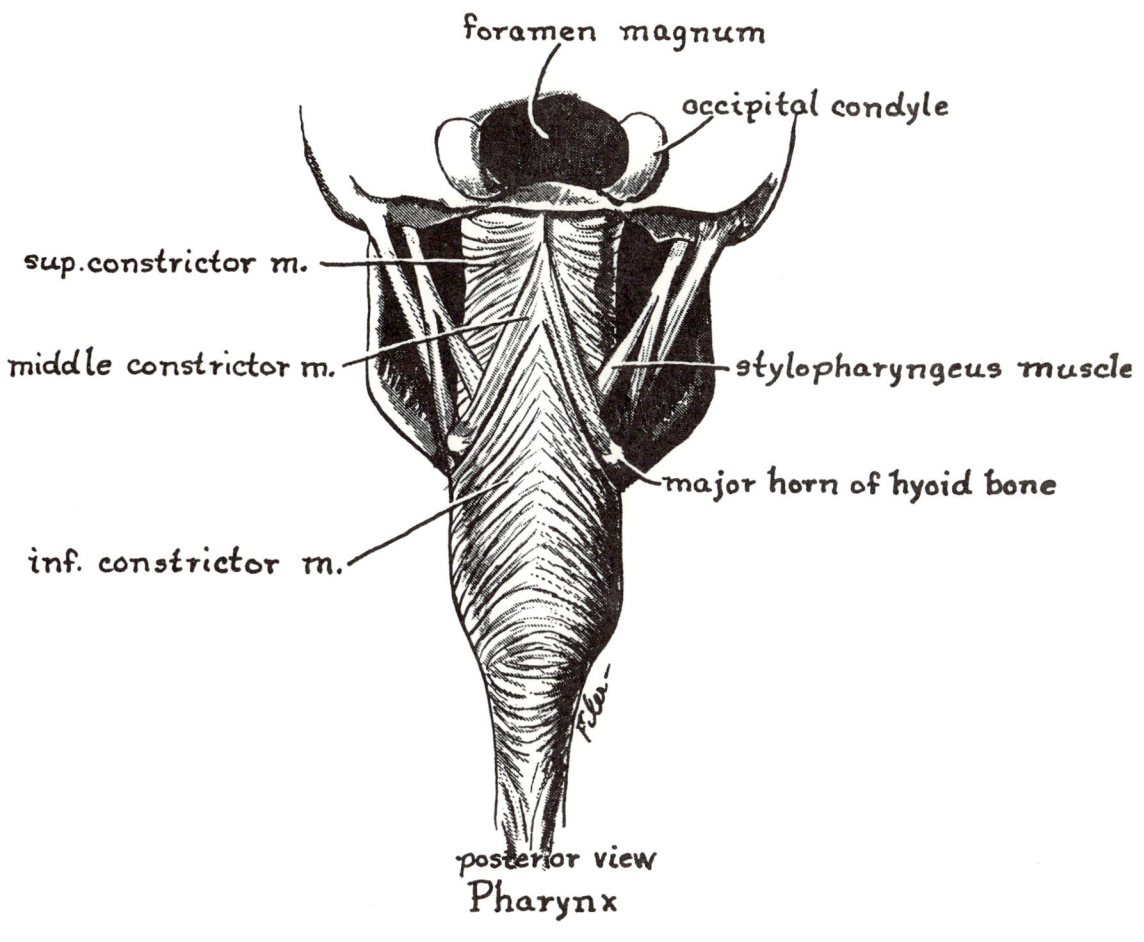

Figure 91.

PALATOPHARYNGEUS. See Muscles of the Soft Palate.

MUSCLES OF THE SOFT PALATE

AZYGOS UVULUS (unpaired). A slender group of muscle fibers extending longitudinally in the midline of the soft palate. (See Fig. 94.)

ATTACHMENTS

To the posterior nasal spine of the palatine bone.

To the uvula of the soft palate.

FUNCTION

Draws the uvula up and shortens the palate.

TENSOR PALATINI (paired). A wide, thin muscle extending vertically from the base of the skull. Becomes tendinous before looping around the hamulus and entering the side of the soft palate.

ATTACHMENTS

To the angular spine and scaphoid fossa of the sphenoid bone and to the lateral wall of the **auditory (Eustachian) tube.**

By a long tendon which loops around the humulus and passes medially into the most anterior part of the soft palate. Attaches to the posterior rim of the hard palate. The tendons from the two sides join to form the **palatal aponeurosis.**

FUNCTION

Opens the orifice of the auditory tube.

LEVATOR PALATINI (paired). A broad, thick muscle immediately inferior to the auditory tube. (See Figs. 92-97.)

ATTACHMENTS

To the petrous portion of the temporal bone. Enters the soft palate at approximately a 45-degree angle from posterior to anterior, also angling medially toward the soft palate. The most posterior fibers extend almost to the uvula, while the most anterior fibers reach into the anterior third of the soft palate. Fibers from the two sides interweave at midline.

FUNCTION

Lifts the soft palate upward and back toward the posterior wall of the pharynx.

Because of its slinglike entrance into the soft palate, it may move the lateral pharyngeal walls medially and posteriorly.

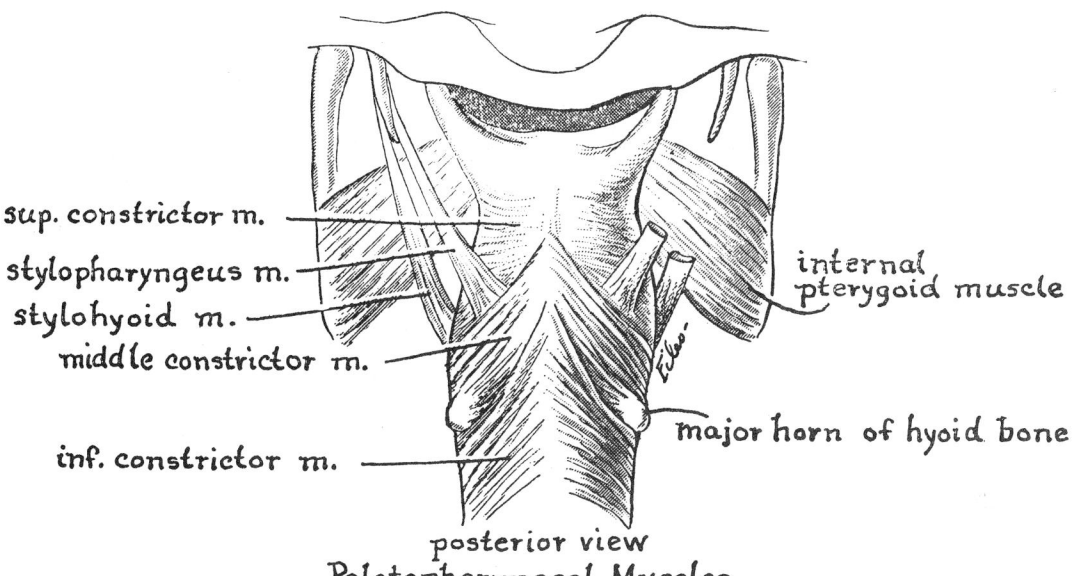

Figure 92.

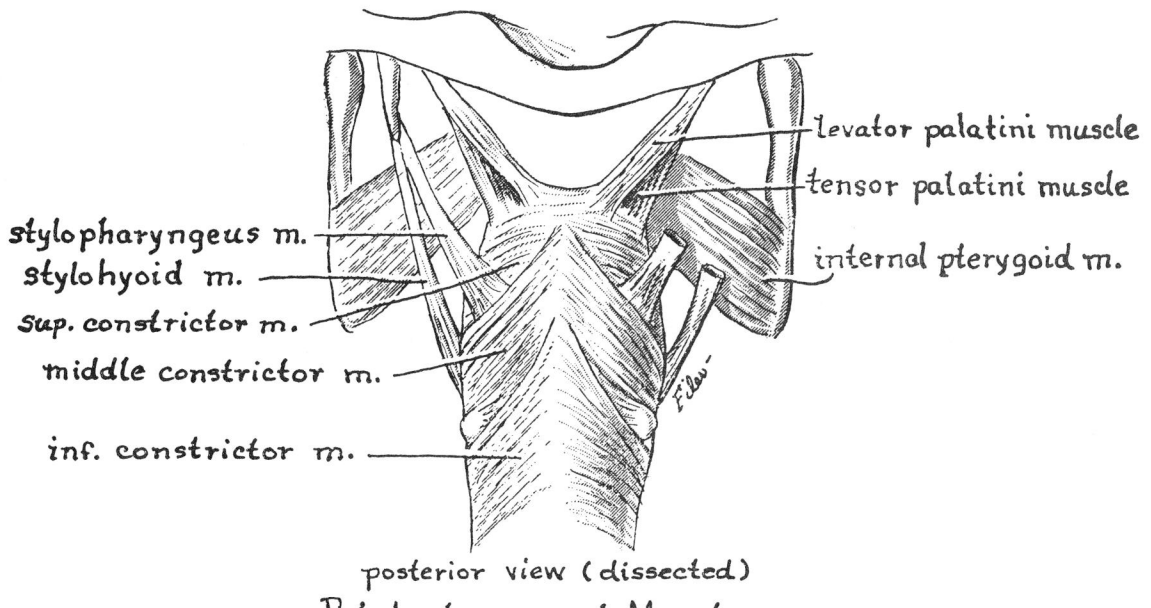

Figure 93.

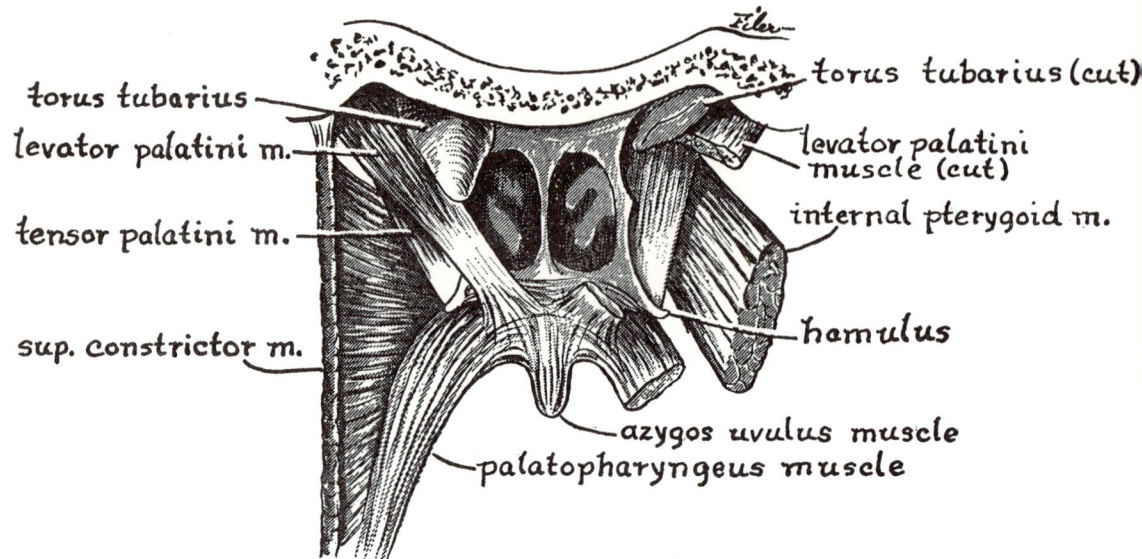

posterior view – dissected
Palatopharyngeal Muscles

Figure 94.

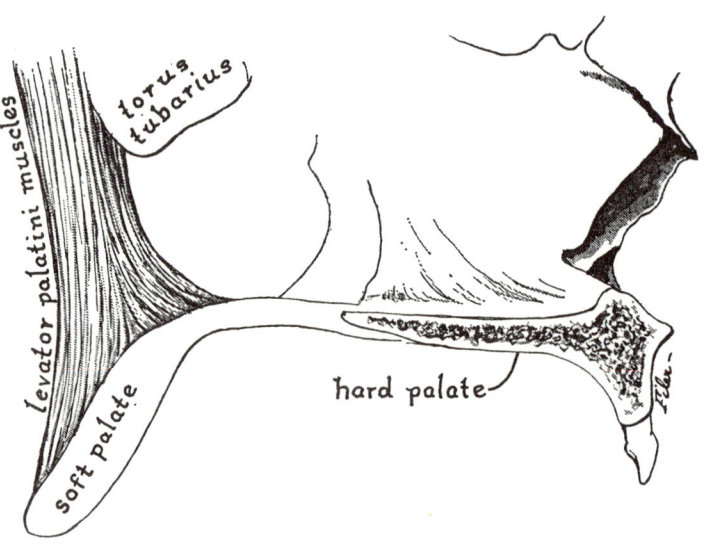

Extent of Insertion of the Levator Palatini Muscle into the Velum

Figure 95.

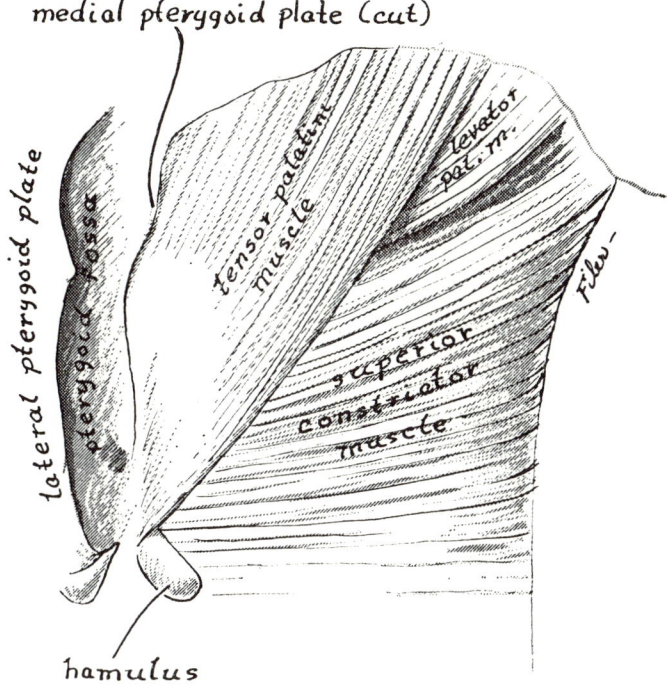

lateral view
Muscles of Velum and Pharynx

Figure 96.

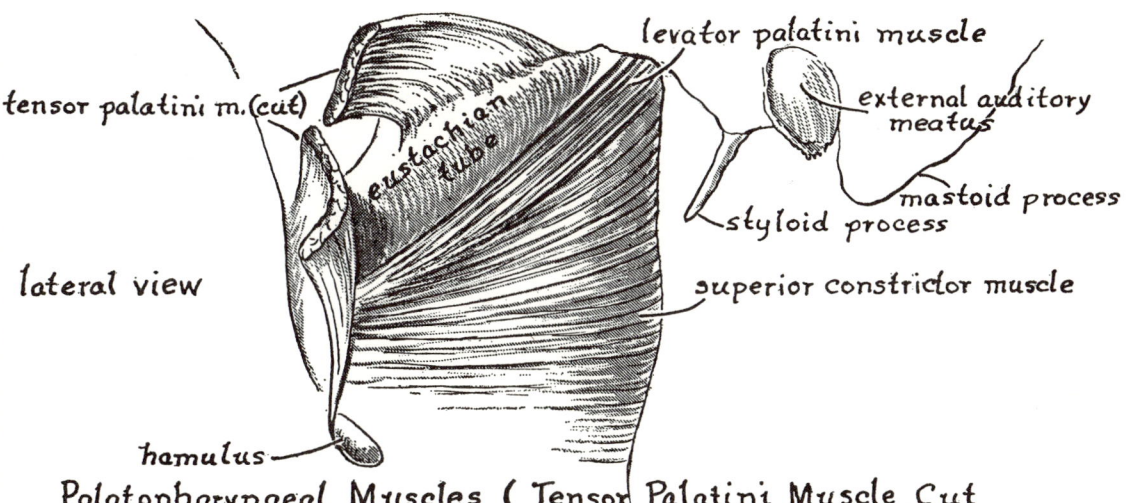

Palatopharyngeal Muscles (Tensor Palatini Muscle Cut to Show its Attachment on Eustachian Tube)

Figure 97.

PALATOPHARYNGEUS (paired). A broad, thick muscle entering the soft palate from the lateral pharyngeal walls. Forms the **posterior pillar of fauces**. (See Figs. 98 and 99.)

ATTACHMENTS

Complex attachment in the pharyngeal wall and to the posterior border of the thyroid cartilage.

One bundle of fibers passes upward to enter the sides of the soft palate. Other fibers pass horizontally into the soft palate from the lateral pharyngeal wall. The anterior bundles from each side join at midline in the soft palate between the tensor and levator muscles.

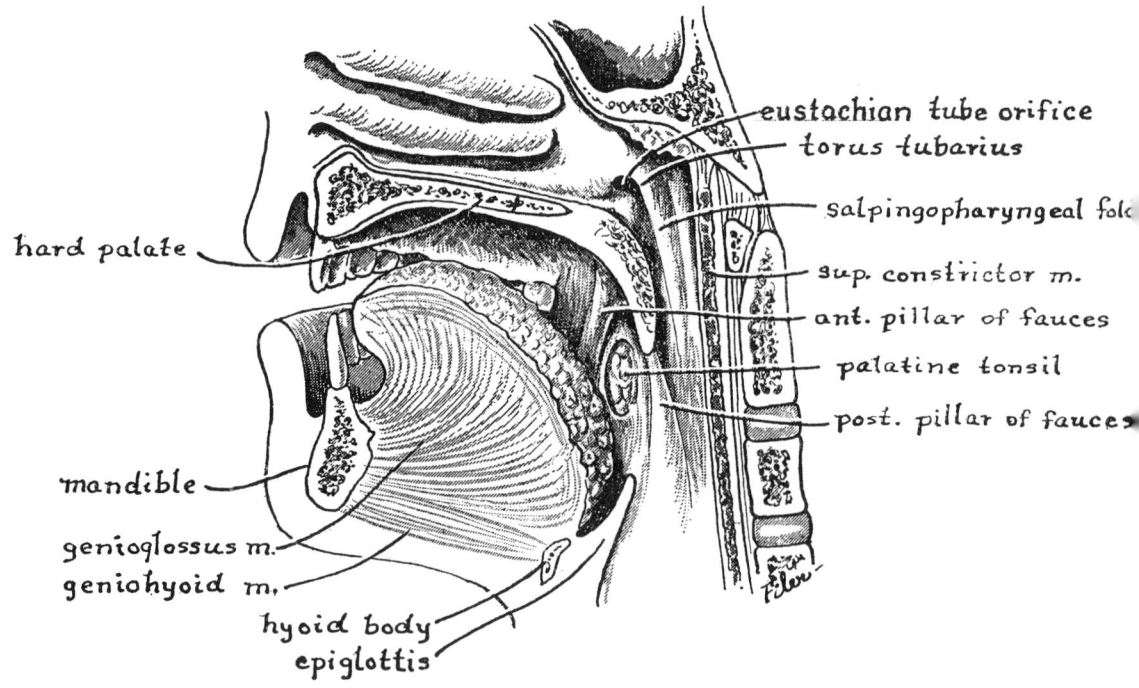

Mid-sagittal Section of Head

Figure 98.

FUNCTION

Raises the lower pharynx and reduces the diameter of the lower pharynx in swallowing.

Lowers the soft palate.

PALATOGLOSSUS (paired). Forms the **anterior pillar of fauces.**

ATTACHMENTS

To the oral surface of the soft palate posteriorly.

To the side of the tongue.

FUNCTION

Pulls the soft palate downward and forward.

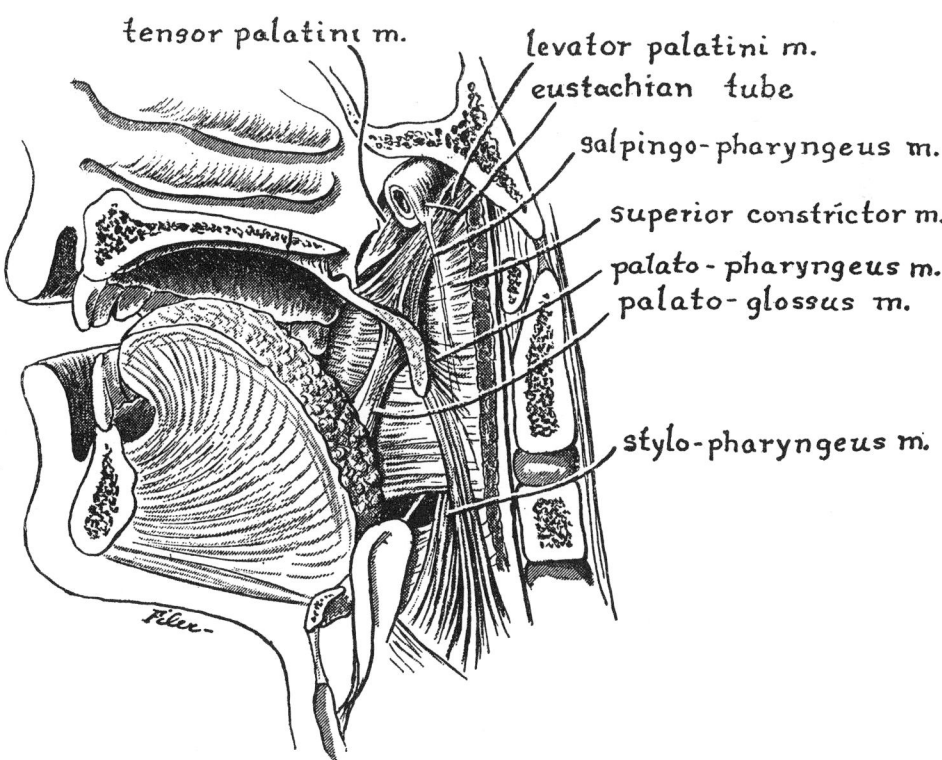

Mid-sagittal Section of Head (dissected)

Figure 99.

May pull the back of the tongue upward and backward.

May move the anterior pillar medially in swallowing.

SALPINGOPHARYNGEUS (paired). A sparse group of muscle fibers in the lateral pharyngeal wall medial to the pharyngeal constrictor muscles. Frequently absent.

ATTACHMENTS

Continuous with the lower fibers of the palatopharyngeus muscle.

To the **torus tubarius** (the cartilage of the auditory tube) or to the lateral wall of the auditory tube.

FUNCTION

May aid in opening the orifice of the auditory tube.

MUSCLES OF THE MANDIBLE

Three muscles which aid in lowering the mandible (the anterior belly of digastric, mylohyoid, and geniohyoid) are described with the extrinsic muscles of the larynx. The mandible lowers in a hingelike action except in extensive openings when the movement is downward and forward. Lateral and anterior-posterior movements occur in mastication.

MASSETER (paired). A thick, powerful muscle composed of a superficial and a deep part. Superficial to the ramus of the mandible. (See Fig. 100.)

ATTACHMENTS

Superficial part to the anterior two thirds of the zygomatic arch. Extends downward and slightly posteriorly to the superficial surface of the angle of the mandible.

Deep part to the length of the zygomatic arch. Extends downward to attach to the lateral surface of the ramus and part of the coronoid process of the mandible. Its deep fibers parallel the superficial fibers except at their most posterior extent, where the deep fibers are vertical.

FUNCTION

Raises the mandible. With one side contracting, may aid in lateralization of the mandible.

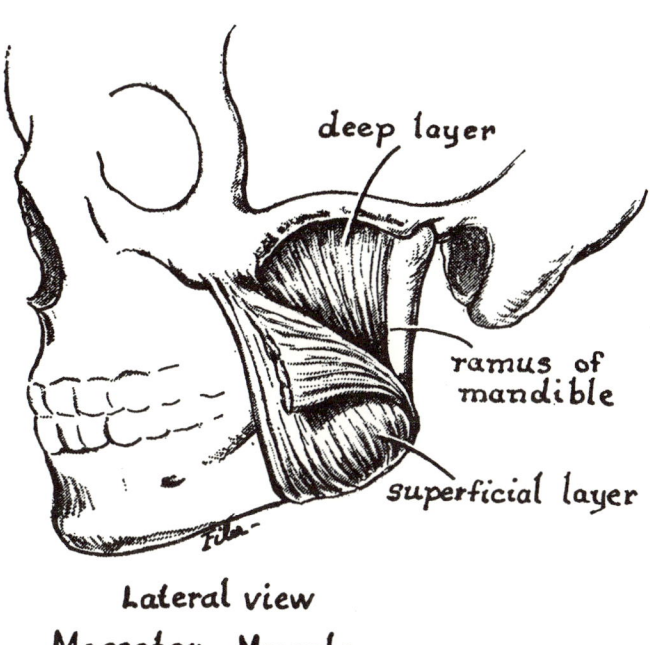

Lateral view
Masseter Muscle

Figure 100.

EXTERNAL PTERYGOID (paired). A two-bellied muscle. The two bellies have a common posterior attachment but diverge slightly anteriorly. (See Fig. 101.)

ATTACHMENTS

To the neck of the mandibular condyle.

To the infratemporal crest and lower margin of the greater wing of the sphenoid bone.

To the lateral surface of the lateral pterygoid plate of the sphenoid bone.

130 Human Vocal Anatomy

FUNCTION

With both sides acting, pulls the mandibular condyles forward. May thus aid in opening the jaws. With one side contracting, may aid in lateralization of the mandible.

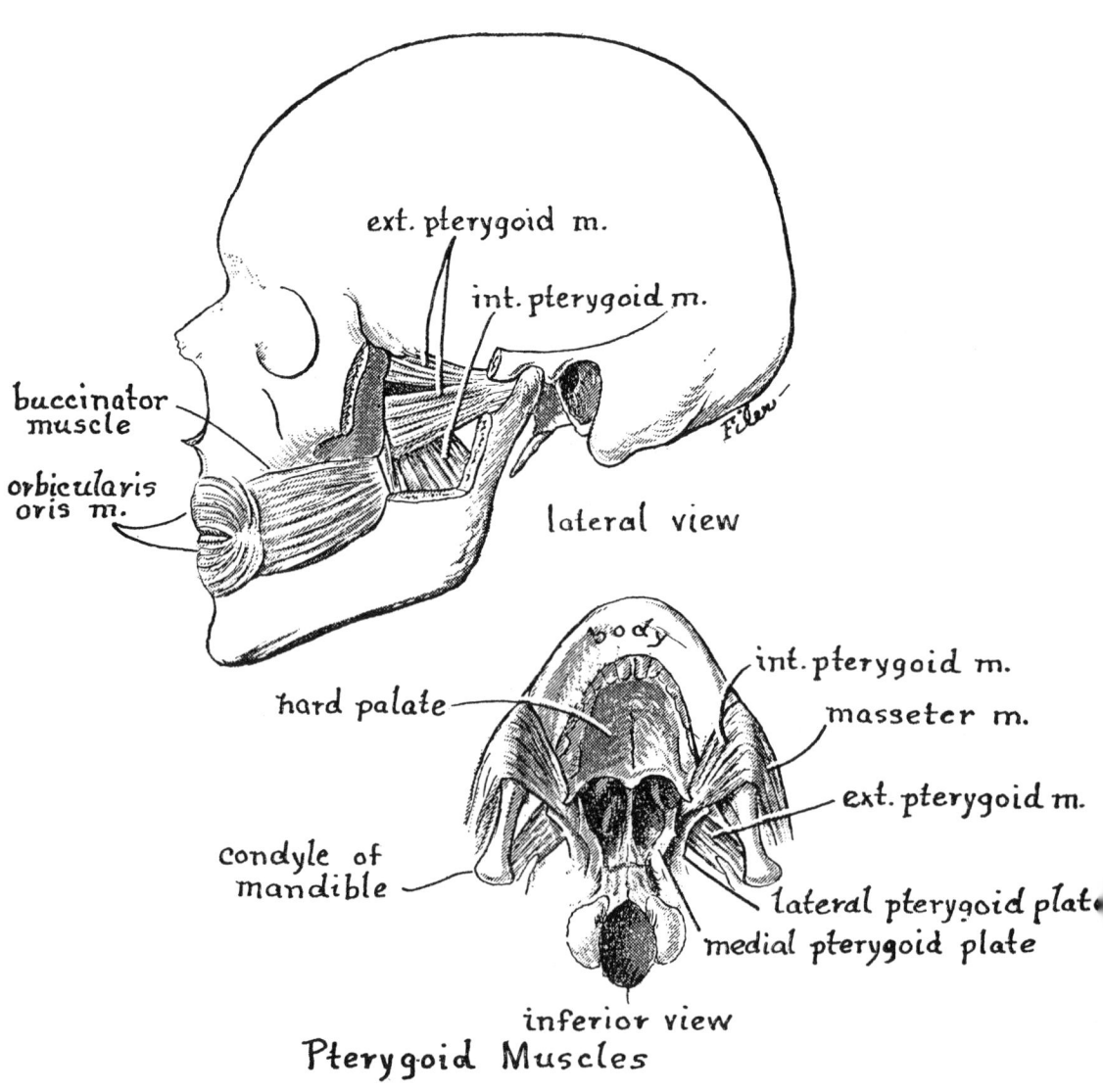

Figure 101.

INTERNAL PTERYGOID (paired). Parallels the masseter muscle but lies deep to the ramus of the mandible. With the masseter, forms a sling around the angle of the mandible.

ATTACHMENTS

To the medial surface of the lateral pterygoid plate of the sphenoid bone and to the pyramidal process of the palatine bone.

Extends downward, posteriorly and laterally to the angle of the mandible.

FUNCTION

Raises the mandible. With one side contracting, may aid in lateralization of the mandible.

TEMPORALIS (paired). Broad and fan-shaped. Located on the lateral surface of the cranium. (See Fig. 102.)

ATTACHMENTS

To the entire length of the inferior temporal line. May attach to the lower part of the greater wing of the sphenoid bone.

To the medial surface and anterior border of the coronoid process of the mandible and to the anterior border of the ramus of the mandible.

FUNCTION

The anterior fibers raise the mandible. The posterior fibers draw the mandible posteriorly. One side contracting may aid in lateralization of the mandible.

MUSCLES OF THE TONGUE

The tongue is composed of an extremely complex array of muscle fibers. While there is general agreement on the gross anatomy of the tongue, the fine detailed anatomy has not been completely described. Extrinsic

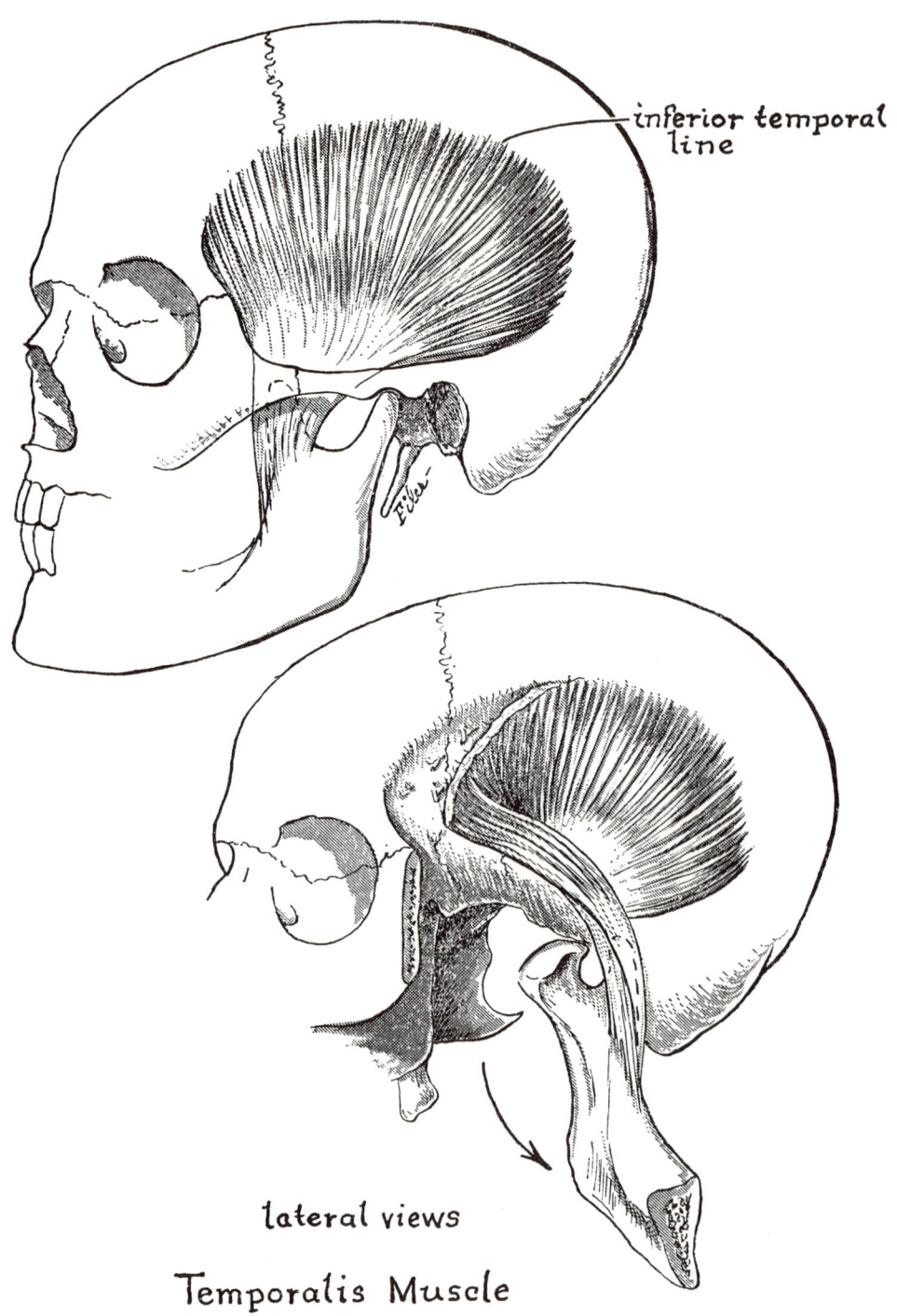

Figure 102.

muscles interweave with intrinsic. Intrinsic muscles interweave with one another, making an accurate and definitive description of their function speculative.

Extrinsic Muscles

STYLOGLOSSUS (paired). The most anterior and superior of the three styloid muscles: styloglossus, stylopharyngeus, stylohyoid.

ATTACHMENTS

To the styloid process of the temporal bone.

Courses downward, forward, and medially into the lateral part of the back of the tongue. Most fibers spread over the lateral part of the tongue. One bundle goes forward to the tip through the lateral margin. Another bundle turns medially at the base of the tongue to penetrate the hyoglossus muscle.

FUNCTION

Draws the tongue upward and backward, especially the back, side, and tip. With one side contracting, may aid in lateralization of the tongue.

PALATOGLOSSUS. See Muscles of the Soft Palate.

HYOGLOSSUS (paired). A thin quadrilateral muscle. (See Fig. 103.)

ATTACHMENTS

To the side of the body and the entire length of the major horn of the hyoid bone.

Courses upward and forward into the root of the tongue at the side medial to the styloglossus muscle and lateral to the inferior longitudinal muscle. Anteriorly its fibers turn forward into the tongue.

FUNCTION

Draws the tongue posteriorly and inferiorly, especially at the side.

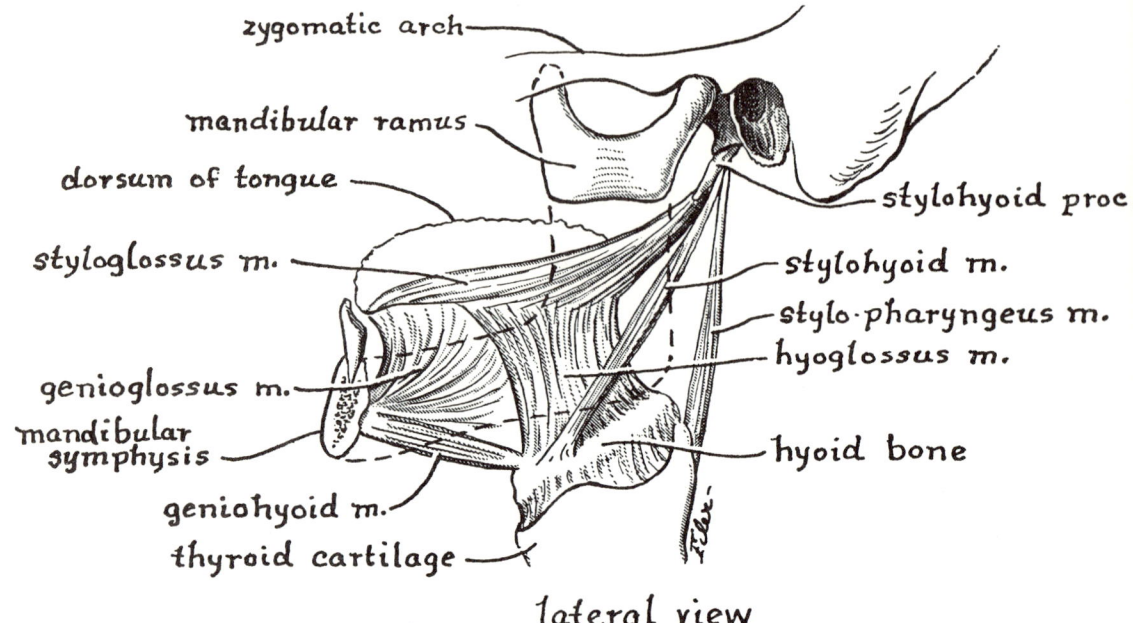

lateral view
Extrinsic Muscles of the Tongue

Figure 103.

GENIOGLOSSUS (paired). A fan-shaped muscle. At midline its two sides are separated over part of their length by a thin layer of connective tissue, **the lingual septum.**

ATTACHMENTS

To the superior mental spine of the mandible.

Its inferior fibers extend to the body of the hyoid bone. The majority of the fibers radiate in the midsagittal plane throughout the length of the tongue, except for the tip.

FUNCTION

Draws the tongue down toward the floor of mouth, especially at midline longitudinally. May draw the hyoid bone and back of the tongue forward to aid in tongue protrusion.

Intrinsic Muscles

SUPERIOR LONGITUDINAL (unpaired). Composed of short fibers coursing largely anteroposteriorly. Covers the entire dorsum of the tongue. (See Fig. 104.)

ATTACHMENTS

Covers the dorsum of the tongue. Fibers extend from the tip to the back of the tongue.

FUNCTION

May shorten the tongue. Fibers acting on one side may aid in tongue lateralization. May make the dorsum of the tongue concave.

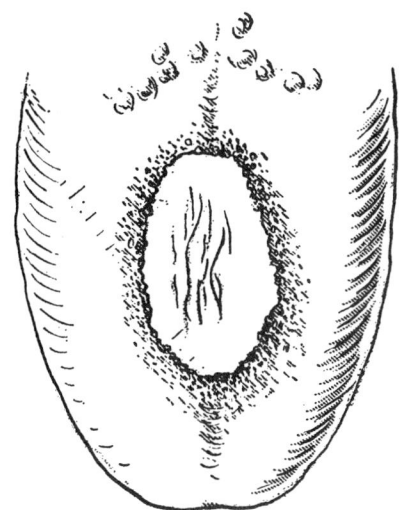

Transverse Section of the Dorsum of the Tongue showing fibers of the Superior Longitudinal Muscle

Figure 104.

INFERIOR LONGITUDINAL (paired). A narrow band of muscle located lateral to the genioglossus muscle.

ATTACHMENTS

To the tip of the tongue.

To the root of the tongue near the minor horns of the hyoid bone.

FUNCTION

May help to shorten the tongue. May depress the tongue tip.

VERTICAL (paired). With the transverse muscle, forms much of the mass of the tongue. (See Figs. 107-108.)

ATTACHMENTS

Its fibers course downward and laterally from the dorsum to the mucous membrane of the inferolateral part of the tongue. These fibers are found mainly anterior to the root of the tongue.

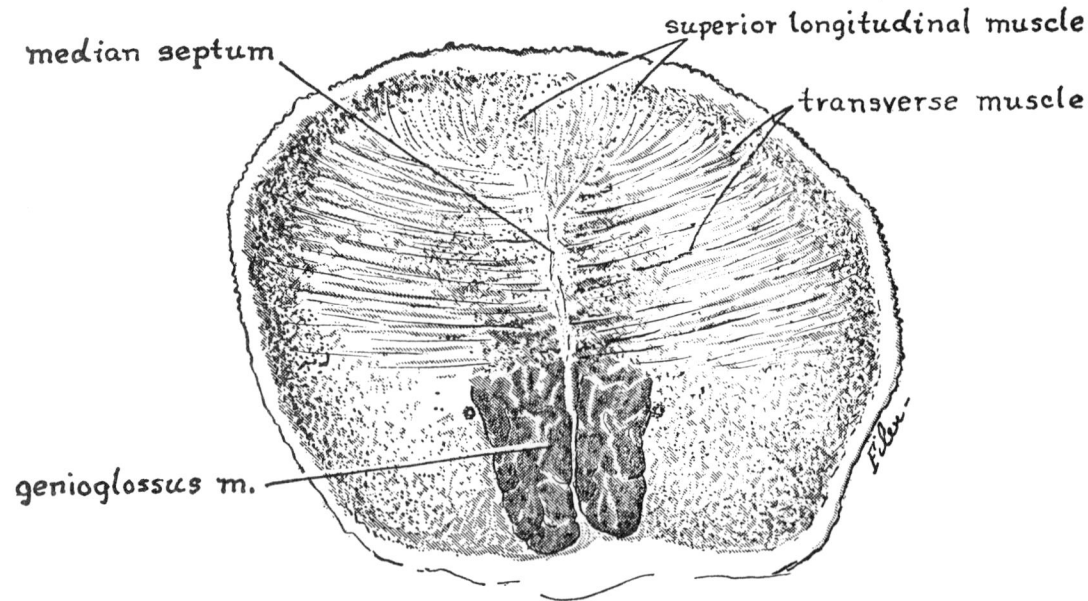

Transverse Section of the Tongue

Figure 105.

FUNCTION

Flattens the tongue. With fibers of the transverse muscle, may account for much of the complex shaping of the tongue.

TRANSVERSE (paired). With the vertical muscle, forms much of the mass of the tongue. (See Figs. 105-108.)

ATTACHMENTS

To the **lingual septum.** Some fibers cross midline.

Fibers radiate laterally, inferiorly, and superiorly to the sides of the tongue.

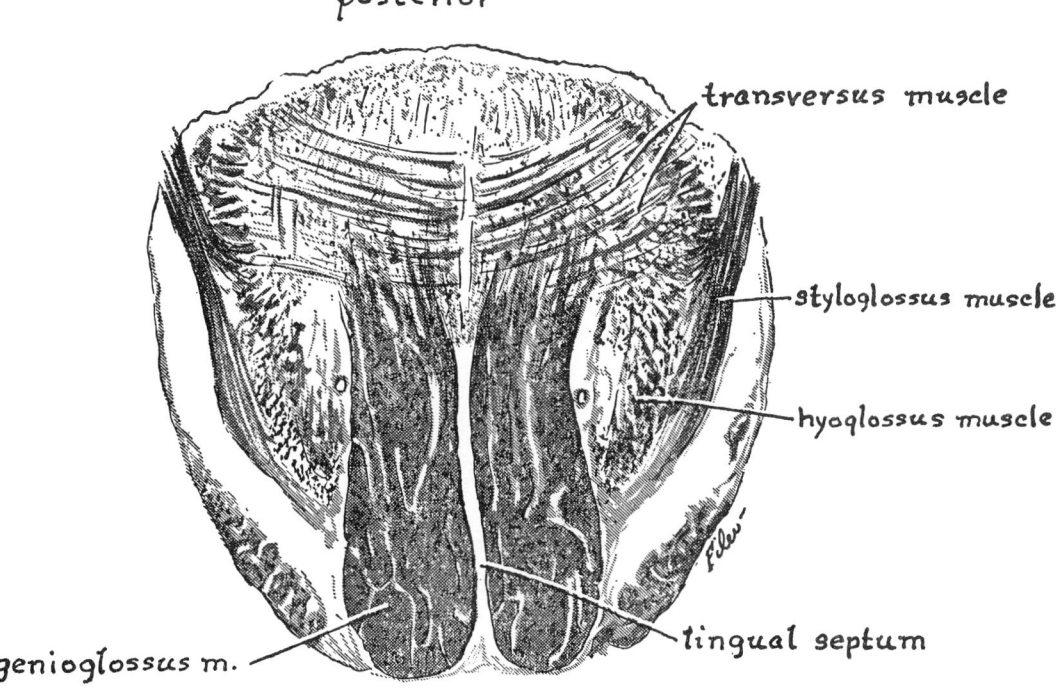

Transverse Section of the Tongue

Figure 106.

FUNCTION

Narrows the tongue. With fibers of the vertical muscle, may account for much of the complex shaping of the tongue.

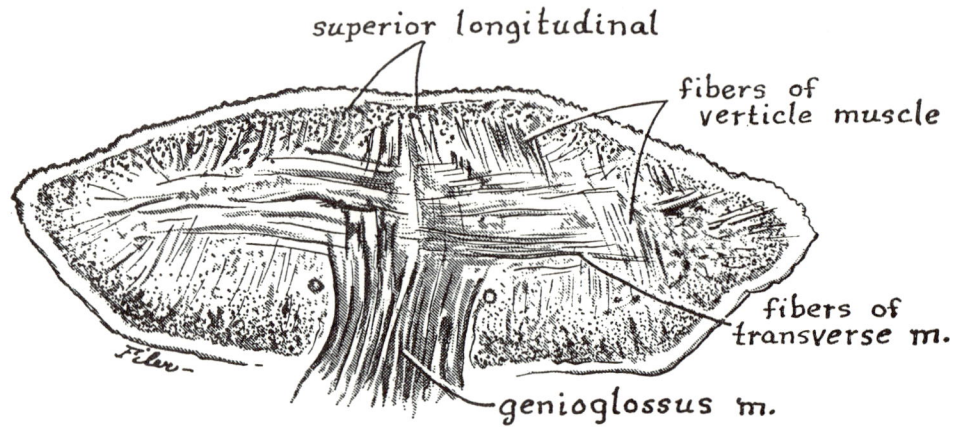

Frontal Section of the Tongue

Figure 107.

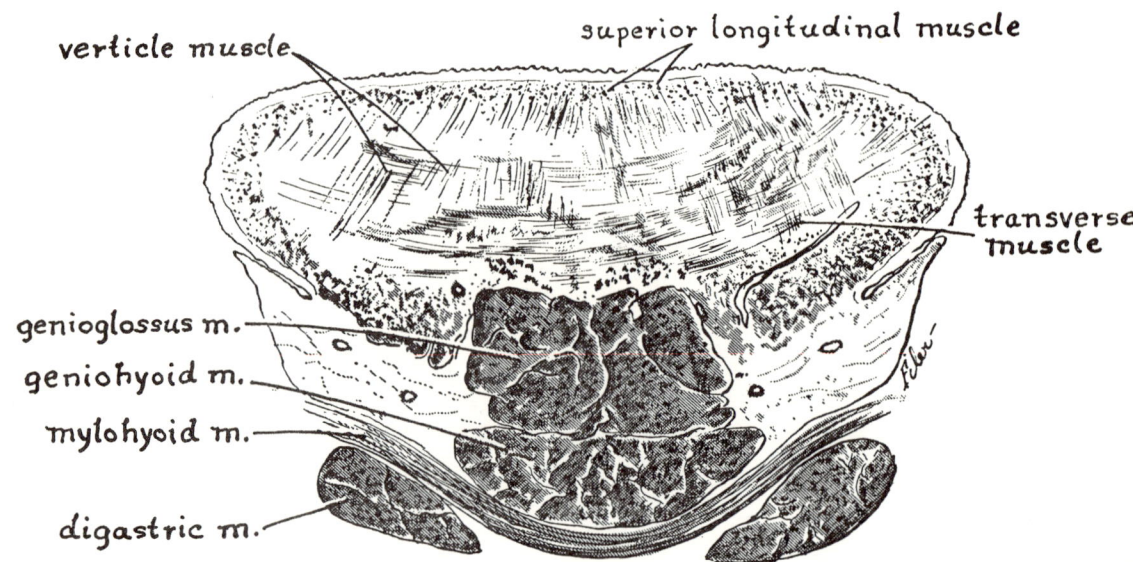

Frontal Section of the Tongue

Figure 108.

MUSCLES OF THE FACE

Only those muscles of the face which are concerned with movements of the lips and so may function in speech articulation will be considered. (See Fig. 109.)

QUADRATUS LABII SUPERIOR (paired). Composed of three parts on each side of midline.

ATTACHMENTS

Infraorbital head to the lower margin of the orbit. *Angular head* to the superior part of the frontal process of the maxilla. *Zygomatic head* to the superficial surface of the body of the zygomatic bone.

All three parts enter the upper lip to intermingle with fibers of the obicularis oris muscle.

FUNCTION

Draws the upper lip upward.

CANINUS (paired). A small muscle, in part deep to the quadratus labii superior muscle.

ATTACHMENTS

To the canine fossa of the maxilla.

To the upper lip at the corner of the mouth. Intermingles with fibers of the obicularis oris muscle.

FUNCTION

May elevate the corner of the mouth.

ZYGOMATICUS (paired). Lies lateral to the quadratus labii superior muscle.

ATTACHMENTS

To the superficial surface of the body of the zygomatic bone lateral

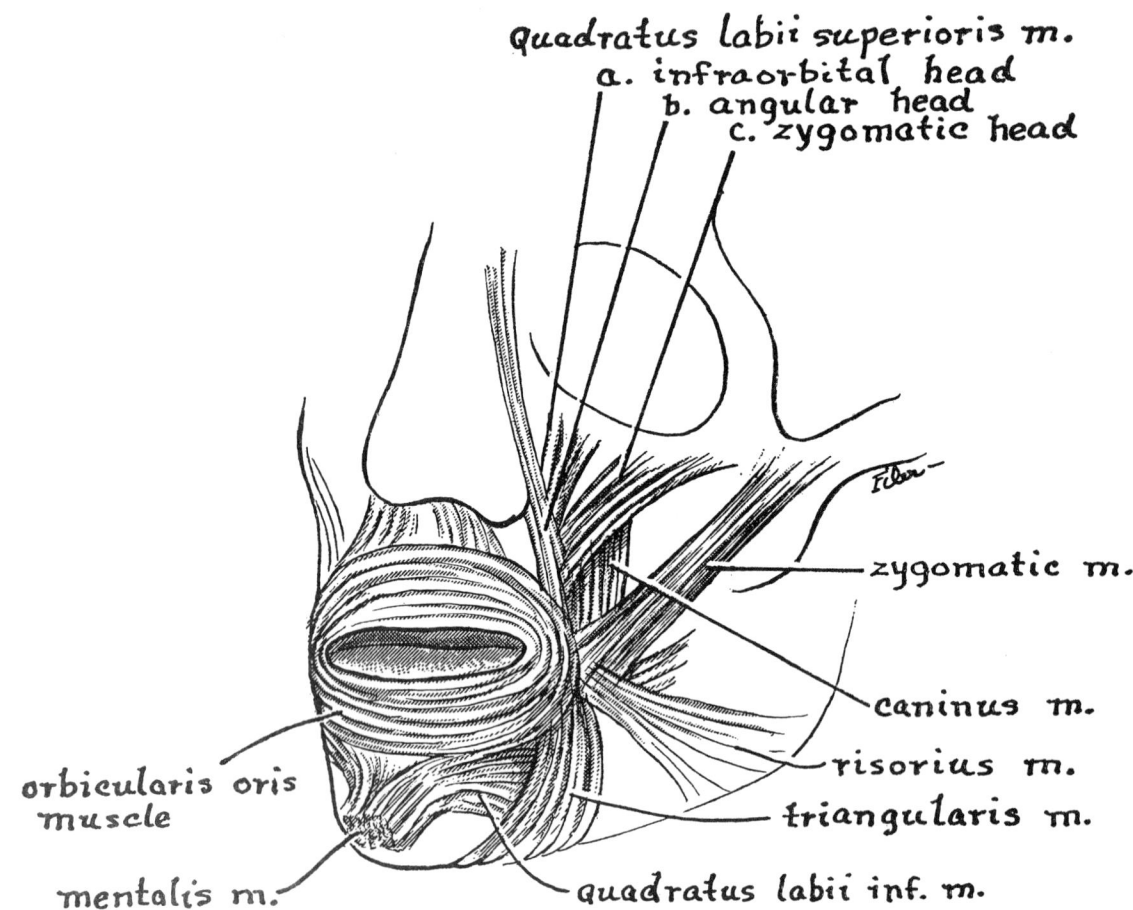

Muscles of the Face

Figure 109.

to the attachment of the zygomatic head of the quadratus labii superior muscle.

To the upper lip at the corner of the mouth. Intermingles with fibers of the obicularis oris muscle.

FUNCTION

Draws the corner of the mouth upward and laterally.

RISORIUS (paired). The most superficial muscle of the lateral part of the face.

ATTACHMENTS

To the fascia, which is superficial to the masseter muscle.

Corner of the mouth. Intermingles with fibers of the obicularis oris muscle.

FUNCTION

Moves the corner of the mouth laterally.

QUADRATUS LABII INFERIOR (paired). A roughly quadrangular muscle of the lower lip.

ATTACHMENTS

To the oblique line of the mandible lateral to its symphysis.

To the lower lip from the midline laterally toward the corner of the mouth. Intermingles with fibers of the obicularis oris muscle.

FUNCTION

Draws the lower lip downward and perhaps slightly laterally.

TRIANGULARIS (paired). A roughly triangular muscle of the lower lip.

ATTACHMENTS

To the oblique line of the mandible, superficial and lateral to the quadratus labii inferior muscle.

To the lower lip at the corner of the mouth. Intermingles with fibers of the obicularis oris muscle. Some of its fibers continue into the upper lip as part of the caninus muscle.

FUNCTION

Draws the corner of the mouth downward and perhaps medially.

MENTALIS (paired). A small conical muscle of the chin.

ATTACHMENTS

To the incisive fossa of the mandible.

To the skin of the chin.

FUNCTION

Raises and protrudes the lower lip.

BUCCINATOR (paired). A large, well-developed muscle of the cheek, deep to other muscles of the face. (See Figs. 89 and 101.)

ATTACHMENTS

To the pterygomandibular raphe and the adjacent superficial surfaces of the alveolar processes of the mandible and maxilla.

Courses forward through the cheek toward the corner of the mouth. Most of its superior fibers enter the upper lip. The most inferior fibers enter the lower lip. Its middle fibers decussate, upper fibers to the lower lip, and lower fibers to the upper lip. The fibers of the buccinator muscle form the deep layer of the obicularis oris muscle.

FUNCTION

May compress the cheek against the teeth.

OBICULARIS ORIS (unpaired). Formed of layers of fibers passing around the oral orifice. The **deep layer** is composed of fibers from the buccinator muscle. The **superficial layer** is largely composed of fibers from the other facial muscles which enter the lips. Fibers intrinsic to the lips are complex. Some pass from superficial to deep. Others attach to the maxilla, mandible, and nasal septum.

FUNCTION

Closes the lips, compresses the lips against the teeth, and protrudes the lips.

PERIPHERAL INNERVATION OF THE MUSCLES OF ARTICULATION

The muscles of the palate and pharynx are, with one exception, supplied by branches of cranial nerves IX and X. The tensor palatini muscle is innervated by cranial nerve V, as are the muscles of the mandible. Muscles of the tongue are supplied by cranial nerve XII, and muscles of the face are supplied by cranial nerve VII.

The efferent nerve fibers to the palate and pharynx come in part from the **pharyngeal plexus.** This plexus is not well understood but seems to be made up in part of branches from the IX and X cranial nerves. The following list of efferent innervation represents what seems to be the consensus in the literature today.

MUSCLE	EFFERENT INNERVATION
Pharynx	
Superior Constrictor	Cranial IX and the pharyngeal plexus.
Middle Constrictor	Cranial IX and/or Cranial X and the pharyngeal plexus.
Inferior Constrictor	Pharyngeal plexus and the superior laryngeal nerve from Cranial XI.
Stylopharyngeus	Cranial IX and the pharyngeal plexus.
Palate	
Azygos Uvulus	Pharyngeal plexus.
Tensor Palatini	Mandibular branch of Cranial V.
Levator Palatini	Pharyngeal plexus.
Palatopharyngeus	Pharyngeal plexus.

| Palatoglossus | Pharyngeal plexus. |
| Salpingopharyngeus | Pharyngeal plexus. |

MUSCLE	EFFERENT INNERVATION
Mandible	
External Pterygoid	Mandibular Branch of Cranial V.
Masseter	Mandibular Branch of Cranial V.
Internal Pterygoid	Mandibular Branch of Cranial V.
Temporalis	Mandibular Branch of Cranial V.
Tongue	
Styloglossus	Cranial XII.
Hyoglossus	Cranial XII.
Genioglossus	Cranial XII.
Superior Longitudinal	Cranial XII.
Inferior Longitudinal	Cranial XII.
Vertical	Cranial XII.
Transverse	Cranial XII.
Face	
Quadratus Labii Superior	Zygomatic Branch of Cranial VII.
Caninus	Zygomatic Branch of Cranial VII.
Zygomaticus	Zygomatic Branch of Cranial VII.
Risorius	Cervicofacial Branch of Cranial VII.
Quadratus Labii Inferior	Cervicofacial Branch of Cranial VII.
Triangularis	Cervicofacial Branch of Cranial VII.
Mentalis	Cervicofacial Branch of Cranial VII.
Buccinator	Cervicofacial Branch of Cranial VII.
Obicularis Oris	Zygomatic Branch and Cervicofacial Branch of Cranial VII.

INDEX

A

Abdominal aponeurosis, 56, 64, 66
Abdominal cavity, 69
Abdominal viscera, 66, 69, 71
Acetabulum of os inominatum, 42
Acromion, of scapula, 48, 73
Alveolar process
  of mandible, 34, 142
  of maxillary bone, 27, 142
Alveoli, 82
Anatomical position, 4
Angle
  inferior
    of occipital bone, 15, 16, 22
    of scapula, 48
  of mandible, 34, 128, 131
  medial, of scapula, 48, 102
  rib, 51, 61, 77, 79, 82
  of thyroid cartilage, 89, 90, 94, 96
Angular head, of quadratus labii superior muscle, 139
Angular spine, of sphenoid bone, 22, 122
Anterior, 4
Anterior belly, of digastric muscle, 99, 100, 115, 128
  innervation of, 115
Anterior cricothyroid ligament, 94
Anterior inferior spine, of ilium, 43
Anterior nasal spine, of maxillary bone, 25
Anterior part, of scalene muscle, 54, 56
Anterior pillar of fauces, 127
Anterior portion, of cricothyroid muscle, 104
Anterior ridge, of arytenoid cartilage, 91
Anterior superior spine, of ilium, 42
Anterior tubercle, of first cervical vertebra, 39, 116
Antero-lateral surface, of arytenoid cartilage, 91, 93
Aorta, 69
Apex, of arytenoid cartilage, 91, 93, 107, 108, 111, 113
Aponeurosis, 3
  abdominal, 56, 64, 66, 69
  palatal, 122
Appendix, of larynx, 86
Arch
  neural, of vertebra, 36

of cricoid cartilage, 87, 88, 94, 104, 106, 109
of zygomatic bone, 25, 31, 128
Arm, 53, 56, 73, 86
Articular facet
  cricoarytenoid, of cricoid cartilage, 88
  crico-thyroid, of cricoid cartilage, 88
  of arytenoid cartilage, 88
  of eleventh and twelfth thoracic vertebra, 41
  of first cervical vertebra, 39
  of typical vertebra, 38
Aryepiglottic fold, of larynx, 86, 93, 96, 113, 114
Aryepiglottic muscle, 109, 113, 115
  function of, 113
  innervation of, 115
  structure of, 113
Arytenoid cartilage, 88, 106, 107, 113, 111, 90, 91, 93, 94, 96, 98, 108, 109
Atlas, first cervical vertebra, 16, 39, 116
Auditory tube, 117, 122, 128
Axis, second cervical vertebra, 40, 40
Azygos uvulus muscle, 121, 122, 143, 124
  function of, 121
  innervation of 143
  structure of 121

B

Basilar portion, of occipital bone, 15, 16
Back, 72, 73
  muscles of, 72
Belly
  anterior, of digastric muscle, 99, 100, 115, 128
  inferior, of omohyoid muscle, 102
  posterior, of digastric muscle, 99, 100, 102, 115
  superior, of omohyoid muscle, 102
  of external pterygoid muscle, 129
Biologic function, 86, 110, 111, 113, 114
Body
  of clavicle, 47
  of eleventh and twelfth thoracic vertebra, 41
  of first cervical vertebra, 39

of first thoracic vertebra, 41
of hyoid bone, 36, 90, 93, 97, 99, 100, 102, 103, 104, 133
of lumbar vertebra, 42
of mandible, 34, 99, 100, 117
of maxillary bone, 25, 27
of ninth thoracic vertebra, 41
of scapula, 48
of second cervical vertebra, 40
of sphenoid bone, 16, 20, 24, 117
of tenth thoracic vertebra, 41
of typical cervical vertebra, 39
of typical thoracic vertebra, 40
of typical vertebra, 36, 51
Body axis, 52
Bone, 3
   atlas, first cervical vertebra, 16, 39, 39, 116
   axis, second cervical vertebra, 40
   clavicle, 46, 47, 53, 56, 59, 73, 102
   ethmoid, 13, 18, 20, 24, 27, 30, 32
   femur, 42
   frontal, 7, 13, 15, 20, 22, 27, 28, 32
   humerous, 73, 48, 56
   hyoid, 7, 36, 90, 93, 97, 100, 102, 103, 104, 116, 119, 133, 134, 136
   inferior conchal, 20, 27, 30, 32, 33
   inferior turbinated, 32, 33
   lacrimal, 13, 20, 27, 28, 33
   mandible, 18, 34, 35, 100, 102, 116, 117, 128, 130, 129, 131, 134, 141, 142, 143, 144
   maxillary, 13, 20, 25, 27, 33, 28, 29, 31, 32, 139, 142, 143
   nasal, 13, 20, 27, 32
   occipital, 15, 16, 18, 24, 39, 53, 72
   os inaminatum, 42, 43, 64, 66, 69, 73, 79, 81
   palatine, 20, 24, 28, 29, 30, 31, 33, 121, 131
   parietal, 15, 16, 18, 22
   rib, 40, 41, 46, 48, 49, 51, 56, 52, 57, 59, 61, 62, 63, 64, 66, 69, 73, 77, 79, 80, 82, 81, 104
      movement of, 51, 52
   sacrum, 36, 42
   scapula, 46, 48, 56, 57, 59, 60, 73, 102
   sphenoid, 13, 15, 16, 18, 20, 21, 22, 24, 30, 31, 116, 117, 122, 129, 131
   sternum, 46, 47, 49, 52, 53, 54, 56, 61, 63, 64, 69, 102, 104
   temporal, 15, 16, 17, 18, 22, 24, 31, 35, 53, 99, 102, 120, 131, 122, 133
   vertebral, 16, 36, 38, 39, 40, 41, 42, 49, 51, 53, 54, 56, 71, 73, 77, 79, 80, 81, 82, 83
   vomer, 20, 24, 27, 30, 31
   zygomatic, 13, 18, 24, 25, 27, 31, 32, 128, 139

Brachial plexus, 83
Bronchi, 82
Buccinator muscle, 142, 144
   function of, 142
   innervation of, 144
   structure of, 142

C

Canine fossa, of maxillary bone, 25, 139
Canine teeth, 27
Caninus muscle, 139, 142, 144
   function of, 139
   innervation of, 144
   structure of, 139
Cartilage, 3
   types of
      hyaline, 3
      elastic, 3
Cartilage
   arytenoid, of larynx, 88, 106, 107, 111, 113, 90, 91, 93, 94, 96, 98, 108, 109
   corniculate, of larynx, 93
   costal, 49, 57, 61, 62, 63, 64, 66, 104
   cricoid, of larynx, 87, 88, 90, 93, 94, 98, 104, 106, 109, 120, 116
   cuneiform, of larynx, 93
   epiglottis, of larynx, 90, 96, 97, 113
   laryngeal, 87, 88, 89, 90, 91, 93
      see also Arytenoid; Corniculate; Cricoid; Cuneiform; Epiglottis; Thyroid
   rib one, 46
   rib two, 46
   rib ten, 46
   thyroid, of larynx, 88, 89, 90, 94, 97, 96, 104, 106, 109, 111, 113, 114, 120, 126
   torus tuberius, of auditory tube, 128
   tracheal, 82, 87
Caudal, 5
Cavity
   abdominal, 69
   glenoid, of scapula, 48
   laryngeal, 116
   nasal, 18, 20, 22, 25, 28, 32, 116
   oral, 100, 116
   orbital, 7, 18, 27, 31, 139
   thoracic, 69
Cell, 3
Central incisor teeth, 27
Central tendon
   of diaphragm, 69
   of omohyoid muscle, 102
Cephalic, 5
Cervical plexus, 83
Cervical vertebra, 16, 36, 39, 39, 40, 54, 53, 56, 73, 77, 79, 82
   atlas, first cervical vertebra, 16, 39, 40, 117
   axis, second cervical vertebra, 40

typical cervical vertebra, 39
Cheek, 31, 142
Chewing, 116, 130
Chin, 142, 54
   see also Mandible
Clavicle, 47, 53, 56, 59, 73, 103
Cleft palate, 116
Coccyx, 36
Cochlea, 18
Column, vertebral, 7, 36, 48, 49, 51, 53, 54, 65, 77, 79, 80, 81, 82
Concha
   medial, of ethmoid bone, 20, 32
   superior, of ethmoid bone, 20
Conchal crest
   of maxillary bone, 25, 27, 33
   of palatine bone, 29, 30, 33
Condyle
   of mandible, 19, 34, 35, 129
   of occipital bone, 15, 39
Condyloid process, of mandible, 19, 35, 129
Connective tissue, 3, 116, 134
   types of
      aponeuroses, 3
      fascia, 3
      ligament, 3
      membrane, 3
      tendon, 3
Conus elasticus, 94, 109
Corniculate cartilage, 93
Cornu
   inferior, of thyroid cartilage, 90, 104
   major, of hyoid bone, 36, 90, 93, 97, 104, 119, 133
   minor, of hyoid bone, 36, 99, 102, 119, 136
   superior, of thyroid cartilage, 90, 93, 97
   see also Horn
Corocoid process, of scapula, 48, 56, 59
Coronal, 4
Coronoid process, of mandible, 35, 128, 131
Corpus of sternum, 46, 47
Costal cartilage, 49, 57, 61, 62, 63, 64
Costal elevators, 79, 80, 85
   function of, 80
   innervation of, 85
   structure of, 79
Costal part, diaphragm, 69
Cranial nerve
   V, 114, 115, 144, 143
   VII, 114, 115, 143, 144
   IX, 143
   X, 114, 115, 143
   XI, 84
   XII, 114, 115, 143, 144
Cranium, 7, 15, 17, 18, 20, 21, 22, 131
Crest
   conchal, of maxillary bone, 25, 27, 33
   conchal, of palatine bone, 29, 30, 33
   ethmoid, of maxillary bone, 25
   ethmoidal, of palatine bone, 20, 29, 30
   infratemporal, of sphenoid bone, 21, 129
   ilium, 42, 66, 69, 73, 79, 81
   pubis, 43, 64, 66, 69
Cribriform plate, of ethmoid bone, 13, 18, 20, 24
Cricoarytenoid articulate facet, of cricoid cartilate, 88
Cricoid cartilage, 87, 88, 90, 93, 94, 98, 104, 106, 109, 116, 120
Cricothyroid articular facet, of cricoid cartilage, 88
Cricoarytenoid ligament, 98
Cricopharyngeus muscle, 120
   see also Inferior constrictor muscle
Cricothyroid membrane, 94
Cricothyroid muscle, 104, 106, 115
   function of, 106
   innervation of, 115
   structure of, 104
Cuneiform cartilage, 93

D

Deep, 5
Deep layer, of obicularis oris muscle, 142
Deep part, of masseter muscle, 128
Demi-facets
   of ninth thoracic vertebra, 41
   of thoracic vertebra, 40
Diaphragm, 66, 69, 82, 83, 84
   function of, 71
   innervation of, 84
   structure of, 69, 71
Digastric muscle, 99, 100, 102, 115, 178
   function of, 100
   innervation of, 115
   structure of, 99
Dorsal, 4
Dorsomedial surface, of arytenoid cartilage, 90, 91, 98
Dorsum, of tongue, 135

E

Ear, 17
Efferent innervation, see Innervation
Elastic cartilage, 3
Epiglottis, 90, 96, 97, 113
Epithelial tissue, 4, 82
   type of
      mucous membrane, 4
      skin, 4
Erect portion, of cricothyroid muscle, 104
Esophagus, 69, 116, 120
Ethmoid bone, 13, 18, 20, 24, 27, 28, 30, 33
Ethmoid crest, of maxillary bone, 25

Ethmoid notch, of frontal bone, 7, 13, 20
Ethmoid process, of inferior conchal bone, 33
Ethmoid crest, of palatine bone, 20, 29, 30
Ethmoidal surface, of palatine bone, 20, 29, 30
Eustachian tube, 116, 117, 122, 128
Exhalation, 53, 61, 62, 63, 65, 66, 69, 73, 77, 79, 82
External abdominal oblique muscle, 62, 66, 84
   function of, 66
   innervation of, 84
   structure of, 66
External auditory meatus, of temporal bone, 17, 18
External intercostal muscles, 61, 84
   function of, 61
   innervation of, 84
   structure of, 61
External occipital protruberance, of occipital bone, 15, 72
External pterygoid muscle, 129, 144
   function of, 130
   innervation of, 144
   structure of, 129
Extrinsic muscles of the larynx, 99, 100, 102, 103, 104, 114, 128
   innervation of, 114, 115
Extrinsic muscles of the tongue, 133, 134
   innervation of, 143, 144

F

Face, 7, 139, 141, 142
   innervation of, 143, 144
   muscles of, 139
Facet
   articular
      of arytenoid cartilage, 93
      of eleventh and twelfth thoracic vertebra, 41
      of first cervical vertebra, 39
      of typical vertebra, 38
   cricoarytenoid articular, of cricoid cartilage, 88
Facet (continued)
   crico-thyroid articular, of cricoid cartilage, 88
   demi-
      of ninth thoracic vertebra, 41
      of thoracic vertebra, 40
   single, of tenth thoracic vertebra, 41
   whole, of thoracic vertebra, 40
Facial surface, of maxillary bone, 25
Fascia, 3
   intercostal, 57, 60
   lumbar, 66, 69, 73, 79
Femur, 42
Floating rib, 49

Fold
   aryepiglottic, of larynx, 66, 86, 93, 96, 113, 114
   ventricular, of larynx, 87, 96, 113
   vocal, of larynx, 86, 100, 109, 113
Foramen
   incisive, of maxillary bone, 27
   magnum, of occipital bone, 15
   transverse, of typical cervical vertebra, 39
   vertebral, of typical vertebra, 36
Foramen magnum, of occipital bone, 15
Fossa
   canine, of maxillary bone, 25, 139
   incisive, of mandible, 34, 142
   inferior, of arytenoid cartilage, 93, 109, 111
   mandibular, of temporal bone, 17
   pterygoid, of sphenoid bone, 22
   scaphoid, of sphenoid bone, 22, 122
   superior, of arytenoid cartilage, 93, 96, 113
Frontal, 4
Frontal bone, 7, 13, 15, 20, 22, 27, 28, 32
Frontal process, of maxillary bone, 13, 25, 27, 28, 32, 139
Frontosphenoidal process, of zygomatic bone, 13, 24, 31

G

Genioglossus muscle, 134, 136, 144
   function of, 134
   innervation of, 144
   structure of, 134
Geniohyoid muscle, 100, 102, 115, 128
   function of, 102
   innervation of, 115
   structure of, 100
Girdle
   pelvic, 7, 36, 42
   shoulder, 7, 46
Glabella, of frontal bone, 7
Glenoid cavity, of scapula, 48
Glottis, of larynx, 86
Greater horn of hyoid bone
   see Major horn and Major cornu
Greater wing, of sphenoid bone, 13, 15, 18, 21, 22, 24, 31, 129, 131

H

Hamulus, of sphenoid bone, 22, 117, 122
Hard palate, 22, 25, 27, 28, 116, 122
Head, 53
Head
   angular, of quadratus labii superior, 139
   infraorbital, of quadratus labii superior, 139
   zygomatic, of quadratus labii superior, 139, 141
   of femur, 42
   of humerous, 48, 56, 73

of rib, 40, 51
of scapula, 73
Heart muscle, 4
Horizontal part, of palatine bone, 28, 30
Horn
    inferior, of thyroid cartilage, 90, 104
    major, of hyoid bone, 36, 90, 93, 97, 104, 119, 133
    minor, of hyoid bone, 36, 99, 102, 119, 136
    superior, of thyroid cartilage, 90, 93, 97
Humerous, 48, 56, 73
Hyaline cartilage, 3
Hyoepiglottic ligament, of epiglottis, 90, 97
Hyoglossus muscle, 133, 144
    function of, 133
    innervation of, 144
    structure of, 133
Hyoid bone, 7, 36, 90, 97, 100, 102, 103, 104, 116, 119, 133, 134, 136

I

Iliac crest, of ilium, 42, 66, 69, 73, 79, 81
Iliocostalis cervicis muscle, 77, 85
    function of, 77
    innervation of, 85
    structure of, 77
Iliocostalis dorsi muscle, 79
    see also Iliocostalis thoracis muscle
Iliocostalis lumborum muscle, 79, 85
    function of, 79
    innervation of, 85
    structure of, 79
Ilicostalis thoracis muscle, 79, 85
    function of, 79
    innervation of, 85
    structure of, 79
Ilium, of os inaminatum, 42, 43
Incisive foramen, of maxillary bone, 27
Incisive fossa, of mandible, 34, 142
Incisor tooth, 34
Inferior, 5
Inferior angle
    of occipital bone, 15, 16, 24
    of scapula, 48
Inferior articular process, of typical vertebra, 38
Inferior belly, of omohyoid muscle, 102
Inferior conchal bone, 20, 27, 30, 32-33
Inferior constrictor muscle, 120, 143
    function of, 120
    innervation of, 143
    structure of, 120
Inferior fossa, of arytenoid cartilage, 93, 109, 111
Inferior horn, of thyroid cartilage, 90, 104
Inferior longitudinal muscle, 133, 136, 144
    function of, 136
    innervation of, 144
    structure of, 136
Inferior mental spine, of mandible, 34, 100
Inferior temporal line
    of frontal bone, 7, 131
    of parietal bone, 15, 131
    of temporal bone, 17, 131
Inferior turbinated bone, 32
Infrahyoid muscles, 102, 103, 104
Infraorbital head, of quadratus labii superior muscle, 139
Infratemporal crest, of sphenoid bone, 21, 129
Infratemporal surface, of maxillary bone, 25
Inguinal ligament, 66, 69
Inhalation, 52, 53, 54, 56, 57, 59, 60, 61, 62, 71, 73, 77, 79
Innervation
    efferent
        of extrinsic laryngeal muscles, 114, 115
        of extrinsic muscles of the tongue, 144
        of intrinsic laryngeal muscles, 114
        of intrinsic muscles of the tongue, 144
        of muscles
            of the abdomen, 84
            of the anterior thorax, 84
            of the back, 84
            of the face, 144
            of the larynx, extrinsic, 114
            of the larynx, intrinsic, 114
            of the mandible, 144
            of the neck, 84
            of the pharynx, 143
            of the soft palate, 143, 144
            of the tongue, extrinsic, 144
            of the tongue, intrinsic, 144
    muscle
        aryepiglottic, 115
        azygos uvulus, 143
        buccinator, 144
        caninus, 144
        costal elevator, 85
        cricopharyngeus, *see* Inferior constrictor muscle
        cricothyroid, 115
        diaphragm, 84
        digastric, 115
        external abdominal oblique, 84
        external intercostals, 84
        external pterygoid, 144
        genioglossus, 144
        geniohyoid, 115
        hyoglossus, 144
        iliocostalis cervicis, 85
        iliocostalis dorsi, *see* Iliocostalis thoracis
        iliocostalis lumborum, 85
        iliocostalis thoracis, 85
        inferior constrictor, 143

inferior longitudinal, 144
internal abdominal oblique, 84
internal intercostals, 84
internal pterygoid, 144
lateral cricoarytenoid, 115
latissimus dorsi, 85
levator palatini, 143
masseter, 144
medial thyroarytenoid, 115
mentalis, 115
middle constrictor, 143
mylohyoid, 115
obicularis oris, 144
oblique arytenoid, 115
omohyoid, 115
palatoglossus, 144
palatopharyngeus, 143
pectoralis major, 84
pectoralis minor, 84
posterior cricoarytenoid, 115
quadratus labii inferior, 144
quadratus labii superior, 144
quadratus lumborum, 85
rectus abdominis, 84
risorius, 144
salpingopharyngeus, 144
scalenes, 84
serratus anterior, 84
serratus posterior inferior, 85
serratus posterior superior, 85
sternocleidomastoid, 84
sternohyoid, 115
sternothyroid, 115
styloglossus, 144
stylohyoid, 115
stylopharyngeus, 143
subclavius, 84
subcostals, 85
superior constrictor, 143
superior longitudinal, 144
temporalis, 144
tensor, palatini, 143
thyroarytenoid, 115
thyroepiglottic, 115
thyrohyoid, 115
transverse, 144
  abdominal, 84
  arytenoid, 115
  thoracis *see* triangularis sterni
  trapezius, 85
  triangularis, 144
  triangularis sterni, 84
  ventricular, 115
  vertical, 86
  vocalis, 115
  zygomaticus, 144
peripheral, muscles of the larynx, 114

Intercostal fascia, 57, 60
Intermediate tendon, digastric muscle, 99, 102
Internal abdominal oblique muscle, 66, 69, 84
  function of, 69
  innervation of, 84
  structure of, 66
Internal intercostal muscles, 61, 62, 84
  function of, 61, 62
  innervation of, 84
  structure of 61
Internal pterygoid muscle, 131, 144
  function of, 131
  innervation of, 144
  structure of, 131
Intrinsic muscles of the larynx, 86, 104
  innervation of, 114, 115
Intrinsic muscles of the tongue, 135
  innervation of, 144
Involuntary muscle, 4
Ischium, of os inaminatum, 42

J

Jaw, 25, 34, 130
Joint, temporomandibular, 35

L

Lacrimal bone, 13, 20, 27-28, 33
Lacrimal process, of inferior conchal bone, 33
Lamina
  papyracea, of ethmoid bone, 18, 20, 27, 28
    of cricoid cartilage, 88, 98, 104
    of thyroid cartilage, 89, 104, 109, 111, 113, 114, 120
    of typical vertebra, 38
Lamina papyracea, of ethmoid bone, 18, 20, 27, 28
Laryngeal cartilages, 87, 88, 89, 90, 91, 93
  see also Arytenoid; Corniculate; Cricoid; Cuneiform; Epiglottis; Thyroid
Laryngeal cavity, 116
Laryngeal ligaments, 93, 98
Laryngopharynx, 116
Larynx, 36, 82, 86, 93, 99, 100, 104, 109, 114, 120, 128
Lateral, 5
Lateral canine tooth, 25
Lateral cricoarytenoid muscle, 109, 115
  function of, 109
  innervation of, 115
  structure of, 109
Lateral incisor teeth, 27
Lateral mass, of ethmoid bone, 18, 20, 30
Lateral portion, of thyroarytenoid muscle, 107, 111, 110
Lateral pterygoid plate, of sphenoid bone, 22, 24, 129, 131

Lateral thyroarytenoid muscle *see* Lateral portion, Thyroarytenoid muscle
Latissimus dorsi muscle, 66, 73, 83, 85
  function of, 73
  innervation of, 85
  structure of, 73
Layer
  deep, of obicularis oris muscle, 142
  superficial, of obicularis oris muscle, 142
Leg, 53
Lesser wing, of sphenoid bone, 13, 22
Levator palatini muscle, 117, 122, 143
  function of, 122
  innervation of, 143
  structure of, 122
Ligaments, 3
  anterior cricothyroid, 94
  cricoarytenoid, of larynx, 98
  hyoepiglottic, of epiglottis, 90, 97
  inguinal, 66
  laryngeal, 93, 94, 96, 97, 98
  nuchal, 72, 77
  posterior cricoarytenoid, of larynx, 98
  posterior cricothyroid, of larynx, 97
  stylohyoid, 119
  thyrohyoid, of larynx, 97
  vocal, of larynx, 94, 109
  ventricular, of larynx, 96
Ligaments of the larynx, 93, 94, 96, 97, 98
Line
  inferior temporal
    of frontal bone, 7, 131
    of parietal bone, 15, 131
    of temporal bone, 17, 131
  mylohyoid, of mandible, 34, 100, 117
  oblique
    of mandible, 34, 141
    of thyroid cartilage, 90, 104
  superior nuchal, of occipital bone, 15, 53, 72
  supramastoid, of temporal bone, 17
Linea alba, 64, 66, 69
Linea semilunaris, 64
Lingual septum, 134, 137
Lips, 116, 139, 141, 142, 143
Location, anatomical terms for, 3, 4, 5
Lumbar fascia, 66, 69, 73, 79
Lumbar part, diaphragm, 71
Lumbar plexus, 83
Lumbar vertebra, 42, 36, 71, 73, 77, 81
Lungs, 82

M

Major cornu, of hyoid bone, 36, 90, 93, 97, 104, 119, 133
  See also Major Horn

Mandible, 18, 34-35, 100, 102, 116, 117, 128, 130, 129, 131, 134, 141, 142, 143, 144
Mandibular fossa, of temporal bone, 17
Mandibular symphysis, 34, 99, 100, 141
Mandibular teeth, 34
Manubrium, of sternum, 46, 47, 53, 102, 104
Margin, orbital, of frontal bone, 7
Masseter muscle, 128, 129, 141, 144
  function of, 129
  innervation of, 144
  structure of, 128
Mastication, 116, 128
Mastoid portion, of temporal bone, 16, 18
Mastoid process, of temporal bone, 18, 53, 99
Maxilla *see* Maxillary bone
Maxillary bone, 13, 20, 25,25, 27, 28, 29, 31, 32, 32, 139, 142, 143
Maxillary process
  of inferior conchal bone, 33
  of zygomatic bone, 32, 27
Maxillary teeth, 27
Maxillary tuberosity, of maxillary bone, 25
Medial, 5
Medial angle, of scapula, 48, 102
Medial concha, of ethmoid bone, 20, 32
Medial part, of scalene muscle, 54, 56
Medial portion, of thyroarytenoid muscle, 111, 113
Medial pterygoid plate, of sphenoid bone, 22, 24, 117, 12
Medial surface, of arytenoid cartilage, 91
Medial thyroarytenoid muscle, 109
  see also Vocalis muscle
Median raphe, of mylohyoid muscle, 100
Median pharyngeal raphe, 117, 120
Membrane, 4
  cricothyroid, of larynx, 94
  laryngeal, 93
  mucous, 4
    of tongue, 136
  thyrohyoid, of larynx, 93, 97
  quadrangular, of larynx, 96
Mental spine
  inferior, of mandible, 34, 100
  superior, of mandible, 34, 134
Mentalis muscle, 142, 144
  function of, 142
  innervation on, 144
  structure of, 142
Middle constrictor muscle, 117, 119, 120, 143
  function of, 119
  innervation of, 143
  structure of, 117, 119
Minor cornu, of hyoid bone, 36, 99, 102, 119, 136
Molar teeth, 34

Mouth, 99, 134, 139, 141, 142
Mucous membrane, 4, 136
Muscle, type of, 3, 4, 53
  involuntary, 4
  smooth, 4
  striated, 3, 4
  voluntary, 3, 4
Muscle
  aryepiglottic, 109, 113, 115
  azygos uvulus, 121, 122, 143
  buccinator, 142, 144
  caninus, 139, 142, 144
  costal elevator, 79, 80, 85
  cricopharyngeus, 120
    see also Inferior constrictor muscle
  cricothyroid, 104, 106, 115
  diaphragm, 66, 69, 82, 83, 84
  digastric, 99, 100, 102, 115, 128
  external abdominal oblique, 62, 66, 84
  external intercostals, 61, 84
  external pterygoid, 129, 144
  extrinsic laryngeal, 114, 115, 128, 99-104
    innervation of, 114, 115
  extrinsic muscles of the tongue, 133, 134, 143, 144
    innervation of, 143, 144
  genioglossus, 134, 136, 144
  geniohyoid, 100, 102, 115, 128
  heart, 4
  hyoglossus, 133, 144
  iliocostalis cervicis, 77, 85
  iliocostalis dorsi see Iliocostalis thoracis
  iliocostalis lumborum, 79, 85
  iliocostalis thoracis, 79, 85
  inferior constrictor, 120, 143
  inferior longitudinal, 133, 136, 144
  infrahyoid, 102, 103, 104
  internal abdominal oblique, 66, 69, 84
  internal intercostals, 61, 62, 84
  internal pterygoid, 131, 144
  intrinsic laryngeal, 86, 104, 105, 106, 107, 108, 109, 110, 111, 112, 113, 114, 115
    innervation of, 114
  intrinsic muscles of the tongue, 135, 144, 143, 136, 137, 138
    innervation of, 144, 143
  lateral cricoarytenoid, 115, 109
  latissimus dorsi, 66, 73, 83, 85
  levator palatini, 117, 122, 143
  masseter, 128, 129, 141, 144
  medial thyroarytenoid, 111, 113
  mentalis, 142, 144
  middle constrictor, 117, 119, 120, 143
  mylohyoid, 100, 115, 128
  obicularis oris, 139, 141, 142, 143, 144
  oblique arytenoid, 115, 108, 109, 113
  omohyoid, 115, 102, 103, 104
  palatoglossus, 133, 127, 128, 144
  palatopharyngeus, 121, 126, 127, 128, 144
  pectoralis major, 56, 83, 84
  pectoralis minor, 56, 57, 61, 83, 84, 73
  posterior cricoarytenoid, 106, 115
  quadratus labii inferior, 141, 144
  quadratus labii superior, 139, 144
  quadratus lumborum, 85, 81, 82
  rectus abdominis, 64, 65, 84
  risorius, 141, 144
  salpingopharyngeus, 120, 128, 144
  scalene, 54, 56, 61, 84
  serratus anterior, 59, 60, 66, 83, 84
  serratus posterior inferior, 77, 85
  serratus posterior superior, 73, 85, 77
  sternocleidomastoid, 53, 54, 84, 99
  sternohyoid, 102, 104, 115
  sternothyroid, 104, 115
  styloglossus, 144, 133
  stylohyoid, 102, 115, 133
  stylopharyngeus, 120, 133, 143
  subclavius, 59, 84
  subcostals, 82, 85
  superior constrictor, 117, 119, 120, 143
  superior longitudinal, 135, 144
  suprahyoid, 99, 100, 102
  temporalis, 131, 144
  tensor palatini, 122, 143
  thyroarytenoid, 110, 111, 113, 115
  thyroepiglottic, 115, 113, 114
  thyrohyoid, 104, 115
  transverse, 137, 138, 144
  transverse abdominal, 69, 83, 84
  transverse arytenoid, 107, 108, 111, 115
  transverse thoracis see Triangularis sterni muscle
  trapezius, 72, 73, 83, 85
  triangularis, 141, 142, 144
  triangularis sterni, 62, 63, 84
  ventricular, 113, 115
  vertical, 136, 137, 144
  vocalis, 115, 109, 110, 113
  zygomaticus, 139, 141, 144
Muscles of the abdomen, 64
  diaphragm, 69, 66, 82, 83, 84
  external abdominal oblique, 66, 62, 84
  innervation of, 84
  internal abdominal oblique, 66, 69, 84
  rectus abdominis, 64, 65, 84
  transverse abdominal, 69, 83, 84
Muscles of the anterior thorax, 56
  external intercostals, 61, 84
  innervation of, 84
  internal intercostals, 61, 62, 84
  pectoralis major, 56, 83, 84
  pectoralis minor, 56, 57, 61, 73, 83, 84

serratus anterior, 59, 60, 66, 83, 84
subclavius, 59, 84
transverse thoracis *see* Triangularis sterni muscle
triangularis sterni, 63, 84
Muscles of the back, 72
  costal elevators, 79, 80, 85
  iliocostalis cervicis, 77, 85
  iliocostalis lumborum, 79, 85
  iliocostalis thoracis, 79, 85
  innervation of, 84, 85
  latissimus dorsi, 66, 73, 83, 85
  quadratus lumborum, 81, 82, 85
  serratus posterior inferior, 77, 85
  serratus posterior superior, 73, 77, 85
  subcostals, 82, 85
  trapezius, 72, 73, 82, 85
Muscles of the face, 139
  buccinator, 142, 144
  caninus, 139, 142, 144
  innervation on, 143, 144
  mentalis, 142, 144
  obicularis oris, 131, 141, 142, 143, 144
  quadratus labii inferior, 141, 144
  quadratus labii superior, 139, 144
  risorius, 141, 144
  triangularis, 141, 142, 144
  zygomaticus, 139, 141, 144
Muscles of the larynx, extrinsic, 99
  infrahyoid, 102
  innervation of, 114
  suprahyoid, 99
Musclos of the larynx, extrinsic
  infrahyoid
    omohyoid, 102, 103, 104
    sternohyoid, 102, 104, 115
    sternothyroid, 104, 115
    thyrohyoid, 104, 115
  suprahyoid
    digastric, 99, 100, 102, 115, 128
    geniohyoid, 100, 102, 115, 128
    mylohyoid, 101, 115, 128
    stylohyoid, 133, 102, 115
Muscles of the larynx, intrinsic, 104
  aryepiglottic, 109, 113, 115
  cricothyroid, 104, 106, 115
  innervation of, 114
  lateral cricoarytenoid, 109, 115
  medial thyroarytenoid, 111, 113
  oblique arytenoid, 115, 108, 109, 113
  posterior cricoarytenoid, 115, 106
  thyroarytenoid, 110, 115, 111, 113
  thyroepiglottic, 113, 114, 115
  transverse arytenoid, 107, 108, 111, 115
  ventricular, 113, 115
  vocalis, 113, 109, 110, 115
Muscles of the mandible, 128
  anterior belly, of digastric, 99, 100, 115, 128
  external pterygoid, 129, 144
  geniohyoid, 100, 102, 115, 128
  innervation of, 143, 144
  internal pterygoid, 131, 144
  masseter, 128, 129, 141, 144
  mylohyoid, 100, 115, 128
  temporalis, 131, 144
Muscles of the neck, 53
  innervation of, 84
  scalenes, 54, 56, 61, 84
  sternocleidomastoid, 53, 54, 84, 99
Muscles of the pharynx, 117
  cricopharyngeus *see* Inferior constrictor muscle
  inferior constrictor, 120, 143
  innervation of, 143
  middle constrictor, 117, 119, 120, 143
  palatopharyngeus, 121, 126, 127, 128, 144
  salpingopharyngeus, 120, 128, 144
  stylopharyngeus, 120, 133, 143
  superior constrictor, 117, 119, 120, 143
Muscles of the soft palate, 121
  azygos uvulus, 121, 122, 143
  innervation of, 143
  levator palatini, 117, 122, 143
  palatoglossus, 127, 128, 133, 144
  palatopharyngeus, 121, 126, 127, 128, 144
  salpingopharyngeus, 120, 128, 144
  tensor palatini, 122, 143
Muscles of the tongue, extrinsic, 131, 133
  genioglossus, 134, 136, 144
  hyoglossus, 133, 144
  innervation of, 143, 144
  palatoglossus, 127, 128, 133, 144
  styloglossus, 144, 133
Muscles of the tongue, intrinsic, 135
  innervation of, 144
  inferior longitudinal, 133, 136, 144
  superior longitudinal, 135, 144
  transverse, 137, 138, 144
  vertical, 136, 137, 144
Muscular process, of arytenoid cartilage, 91, 93, 106, 107, 108, 109, 111
Mylohyoid line, of mandible, 34, 100, 117
Mylohyoid muscle, 100, 115, 128
  function of, 100
  innervation of, 115
  structure of, 100

N

Nasal bone, 13, 20, 27, 32
Nasal cavity, 18, 20, 22, 25, 28, 32, 116
Nasal notch, of maxillary bone, 25
Nasal septum, 18, 30, 143
Nasal spine

anterior, of maxillary bone, 25
  posterior, of palatine bone, 28, 121
Nasal surface
  of lacrimal bone, 28
  of maxillary bone, 25, 27, 29
Nasopharynx, 116
Neck, 36, 53, 54, 82, 86, 102
Neck, of rib, 51
Nerve
  afferent, 4
  cranial
    V, 114, 115, 144, 143
    VII, 115, 143, 144
    IX, 143
    X, 114, 115, 143
    XI, 84
    XII, 114, 115, 143, 144
  efferent, 4
  spinal accessory, 82
Nervous tissue, 4
  type of
    afferent, 4
    efferent, 4
Neural arch, of typical vertebra, 24
Notch
  ethmoid, of frontal bone, 7, 13, 20
  nasal, of maxillary bone, 25
  of thyroid cartilage, 90
Nuchal ligament, 72, 77

O

Obicularis oris muscle, 139, 141, 142, 143, 144
  function of, 143
  innervation of, 144
  structure of, 142, 143
Oblique arytenoid muscle, 108, 109, 113, 115
  function of, 109
  innervation of, 115
  structure of, 108, 109
Oblique line
  of mandible, 34, 141
  of thyroid cartilage, 90, 104
Oblique portion, of cricothyroid muscle, 104
Occipital bone, 15, 16, 18, 24, 39, 53, 72
Occipital condyle, of occipital bone, 15, 39
Odontoid process, of atlas, first cervical vertebrae, 39
Omohyoid muscle, 102, 103, 104, 115
  function of, 104
  innervation, 115
  structure of, 102, 103
Oral cavity, 100, 116
Orbital cavity, 7, 18, 27, 31-32, 139
Orbital margin, of frontal bone, 7
Orbital plate, of frontal bone, 7, 13
Orbital process

of maxillary bone, 26
of palatine bone, 28
of zygomatic bone, 27, 31
Orbital surface
  of lacrimal bone, 28
  of maxillary bone, 20, 25, 27, 32
  of sphenoid bone, 21, 24
Organ, 3
Organ system, 3
Orientation, anatomical terms for, 4, 5
Oropharynx, 116
Os inaminatum, 42, 43, 64, 66, 69, 73, 79, 81

P

Palatal aponeurosis, 122
Palate
  cleft, 116
  hard, 22, 25, 27, 28, 28, 116
  innervation of, 143
  soft, 116, 117, 120, 121, 122, 126, 127
Palatine bone, 20, 24, 28, 29, 30, 31, 33, 121, 131
Palatine process, of maxillary bone, 25, 27, 27, 29
Palatoglossus muscle, 127, 128, 133, 144
  function on, 127, 128
  innervation of, 144
  structure of, 127
Palatopharyngeus muscle, 121, 126, 127, 128, 144
  function of, 127
  innervation of, 144
  structure of, 126
Parietal bone, 15, 16, 18, 22
Part
  anterior, of scalene muscle, 54, 56
  costal, of diaphragm, 69
  deep, of masseter muscle, 128
  horizontal, of palatine bone, 28, 28, 30
  lumbar, of diaphragm, 71
  medial, of scalene muscle, 54, 56
  posterior, of scalene muscle, 54, 56
  sternal, of diaphragm, 69
  superficial, of masseter muscle, 128
  vertical, of palatine bone, 29
Passavant's pad, 117
Pectoralis major muscle, 56, 83, 84
  function of, 56
  innervation of, 84
  structure of, 56
Pectoralis minor muscle, 56, 57, 61, 73, 83, 84
  function of, 57
  innervation of, 84
  structure of, 56, 57
Pedicle, of typical vertebra, 38

Pelvic girdle, 7, 36, 42
Pelvis, 36, 81
Peripheral innervation, of muscles of larynx, 114
Perpendicular plate, of ethmoid bone, 18, 20, 24, 30, 32
Petrosal process, of sphenoid bone, 21, 22, 24
Petrous portion, of temporal bone, 16, 18, 22, 24, 122
Pharyngeal constrictor muscles, 128
  see also Inferior constrictor; Middle constrictor; Superior constrictor
Pharyngeal plexus, 143
Pharynx, 20, 89, 116, 117, 119, 120, 122, 127
  innervation of, 143, 144
Phonation, 111, 113
Pillar of fauces
  anterior, 127
  posterior, 126
Plate
  cribriform, of ethmoid bone, 13, 20, 24
  lateral pterygoid, of sphenoid bone, 22, 22, 24, 129, 131
  medial pterygoid, of sphenoid bone, 22, 24, 117
  orbital, of frontal bone, 7, 13
  perpendicular, of ethmoid bone, 18, 20, 20, 30, 32
Plexus, 83
  brachial, 83
  cervical, 83
  lumbar, 83
  pharyngeal, 143
Portion
  anterior, of cricothyroid muscle, 104
  basilar, of occipital bone, 15, 16
  erect, of cricothyroid muscle, 104
  lateral, of thyroarytenoid muscle, 107, 111, 113
  mastoid, of temporal bone, 16, 18
  medial, of thyroarytenoid muscle, 111, 113
  oblique, of cricothyroid muscle, 104
  petrous, of temporal bone, 16, 18, 22, 24, 122
  posterior, of cricothyroid muscle, 104
  squamous, of temporal bone, 15, 17, 22, 24
  tympanic, of temporal bone, 18
Posterior, 4
Posterior belly, of digastric muscle, 99, 100, 102, 115
  innervation of, 115
Posterior cricoarytenoid ligament, 97
Posterior cricoarytenoid muscle, 115, 106
  function of, 106
  innervation of, 115
  structure of, 106
Posterior nasal spine, of palatine bone, 28, 121

Posterior part, of scalene muscle, 54, 56
Posterior pillar of fauces, 126
Posterior portion, of cricothyroid muscle, 104
Posterior ridge
  of arytenoid cartilage, 91, 107, 114
  of cricoid cartilage, 88, 106
Posterior superior spine, of ilium, 43
Posterior tubercle, of first cervical vertebra, 39
Premaxilla, of maxillary bone, 27
Process
  alveolar, of mandible, 34, 142
    of maxillary bone, 27, 142
  condyloid, of mandible, 18, 34, 129
  corocoid, of scapula, 48, 56, 59
  coronoid, of mandible, 35, 128, 131
  ethmoid, of inferior conchal bone, 33
  frontal, of maxillary bone, 13, 25, 25, 27, 28, 32, 139
  frontosphenoidal, of zygomatic bone, 13, 24, 31
  inferior articular, of typical vertebra, 38
  lacrimal, of inferior conchal bone, 31
  mastoid, of temporal bone, 18, 53, 99
  maxillary
    of inferior conchal bone, 33
    of zygomatic bone, 27
  muscular, of arytenoid cartilage, 91, 93, 106, 107, 108, 109, 111
  odontoid, of atlas, first cervical vertebra, 39
  orbital
    of palatine bone, 28, 29
    of zygomatic bone, 27, 31
  palatine, of maxillary bone, 25, 27, 29, 31
  petrosal, of sphenoid bone, 21, 22, 24
  pterygoid, of sphenoid bone, 22, 22, 30
  pyramidal, of palatine bone, 24, 29, 30, 131
  sphenoid, of palatine bone, 24, 29, 30
  spinous
    of cervical vertebra, 38, 73, 77
    of first cervical vertebra, 39
    of typical vertebra, 39, 77
    of typical thoracic vertebra, 40, 77, 73
  styloid, of temporal bone, 18, 102, 120, 133
  superior articular
    of first cervical vertebra, 39
    of typical vertebra, 38
  temporal, of zygomatic bone, 18, 30
  transverse
    of eleventh and twelfth thoracic vertebra, 41
    of first thoracic vertebra, 40
    of lumbar vertebra, 81
    of typical cervical vertebra, 39, 56, 77, 79
    of typical thoracic vertebra, 40, 49, 51, 79
    of typical vertebra, 38

uncinate, of ethmoid bone, 20, 33
vocal, of arytenoid cartilage, 91, 94, 109
xiphoid, of sternum, 47, 69
zygomatic
   of frontal bone, 7, 13, 31
   of maxillary bone, 25, 32, 27
   of temporal bone, 17, 18, 32
Processes, of lumbar vertebra, 42
Pterygoid fossa, of sphenoid bone, 22
Pterygoid process, of sphenoid bone, 22, 22, 30
Pterygomandibular raphe, 117, 142
Pubic crest, 43, 64, 66, 69
Pubic symphysis, 43
Pubis, os inaminatum, 42, 43, 64
Pyramidal process, of palatine bone, 24, 29, 30, 131
Pyriform sinus, 96, 116

Q

Quadrangular membrane, 96
Quadratus labii inferior muscle, 141, 144
   function of, 141
   innervation of, 144
   structure of, 141
Quadratus labii superior muscle, 139, 144
   function of, 139
   innervation of, 144
   structure of, 139
Quadratus lumborum muscle, 81, 82, 85
   function of, 82
   innervation on, 85
   structure of, 81

R

Ramus, of mandible, 34, 35, 128, 131
Raphe
   median, of mylohyoid muscle, 100
   median pharyngeal, 117, 120
   pterygomandibular, 117, 142
Rectus abdominis muscle, 64, 65, 84
   function of, 65
   innervation of, 84
   structure of, 64
Rectus sheath, of rectus abdominis muscle, 64
Respiration, 51, 52, 53, 54, 85
Rib, 40, 41, 46, 48, 49, 51, 56, 52, 57, 59, 61, 62, 63, 64, 66, 69, 73, 77, 79, 80, 81, 82, 104
   movement of, 51, 52
   rib one, 41, 46, 49, 56, 59, 60, 79, 104
   rib two, 41, 46, 49, 56, 60, 77, 79
   rib three, 49, 57, 60, 77, 79
   rib four, 49, 57, 60, 77, 79
   rib five, 49, 57, 60, 64, 66, 77, 79
   rib six, 49, 60, 64, 66, 77, 79
   rib seven, 49, 60, 64, 66, 69, 79
   rib eight, 49, 60, 66, 69, 79
   rib nine, 41, 49, 60, 66, 69, 73, 77, 79
   rib ten, 41, 46, 49, 66, 69, 73, 77, 79
   rib eleven, 49, 66, 69, 73, 77, 79
   rib twelve, 49, 66, 69, 70, 71, 73, 77, 79, 81
Rib cage, 36, 52, 59, 65, 66, 69, 82
Ridge
   anterior, of arytenoid cartilage, 91
   posterior, of arytenoid cartilage, 91, 107, 114
   posterior, of cricoid cartilage, 88, 106
Risorius muscle, 141, 144
   function of, 141
   innervation of, 144
   structure of, 141
Rostral, 5
Rudimentary tail, 36

S

Sacral vertebra, 36, 73
Sacrospinal muscle group, 77, 79
Sacrum, 36, 42
Sagittal, 4, 5
Salpingopharyngeus muscle, 120, 128, 144
   function on, 128
   innervation on, 144
   structure of, 128
Scalene muscles, 54, 56, 61, 84
   function on, 56
   innervation on, 84
   structure of, 54
Scaphoid fossa, of sphenoid bone, 22, 122
Scapula, 46, 48, 56, 57, 59, 73, 102
Scapular spine, 48, 73
Second premolar tooth, 34
Sella turcica, of sphenoid bone, 20, 12
Septum
   lingual, 134, 137
   nasal, 18, 30, 143
Serratus anterior muscle, 59, 60, 66, 83, 84
   function on, 60
   innervation on, 84
   structure of, 59, 60
Serratus posterior inferior muscle, 77, 85
   function of, 77
   innervation of, 85
   structure of, 77
Serratus posterior superior muscle, 73, 77, 85
   structure of, 73
   function of, 77
   innervation of, 85
Shaft, of rib, 51
Shoulder, 59, 73, 86
Shoulder girdle, 7, 46
Single face, of tenth thoracic vertebra, 41
Sinus, pyriform of larynx, 96, 116

Skeleton, 7, 18
  parts of, 7-52
Skin, 4
  chin, 142
Skull, 7, 122
Smooth muscle, 4
Soft palate, 116, 117, 120, 121, 122, 126, 127
Sphenoid bone, 13, 15, 16, 18, 20, 21, 22, 24, 30, 31, 116, 117, 122, 129, 131
Sphenoid process, of palatine bone, 24, 29, 30
Spinal accessory nerve, 82
Spinal cord, 83, 84
Spine
  angular, of sphenoid bone, 22, 122
  anterior inferior, of ilium, 43
  anterior nasal, of maxillary bone, 25
  anterior superior, of ilium, 42
  inferior mental, of mandible, 34, 100
  posterior nasal, of palatine bone, 28, 121
  posterior superior, of ilium, 43
  superior mental, of mandible, 34, 134
  of scapula, 48, 73
Spine of scapula, 48, 73
Spinous process
  of cervical vertebra, 39, 73, 77
  of first cervical vertebra, 39
  of typical thoracic vertebra, 40, 73, 77
  of typical vertebra, 38, 77
Squamous portion, of temporal bone, 15, 17, 22, 24
Sternal part, of diaphragm, 69
Sternocleidomastoid muscle, 53, 54, 84, 99
  function of, 53, 54
  innervation of, 84
  structure of, 53
Sternohyoid muscle, 102, 104, 115
  function of, 102
  innervation of, 115
  structure of, 102
Sternothyroid muscle, 104, 115
  function of, 104
  innervation of, 115
  structure of, 104
Sternum, 46, 47, 49, 52, 53, 54, 56, 61, 63, 64, 69, 102, 104
Striated muscle, 3, 4
Styloglossus muscle, 144, 133
  function of, 133
  innervation of, 144
  structure of, 133
Stylohyoid ligament, 119
Stylohyoid muscle, 102, 115, 133
  function of, 102
  innervation of, 115
  structure of, 102
Styloid process, of temporal bone, 18, 133, 120, 102

Stylopharyngeus muscle, 120, 133, 143
  function of, 120
  innervation of, 143
  structure of, 120
Subclavius muscle, 59, 84
  function of, 59
  innervation of, 84
  structure of, 59
Subcostal muscles, 82, 85
  function of, 82
  innervation of, 85
  structure of, 82
Superficial, 5
Superficial layer, of obicularis oris muscle, 142
Superficial part, of masseter muscle, 128
Superior, 5
Superior articular process
  of first cervical vertebra, 39
  of typical vertebra, 38
Superior belly, of omohyoid muscle, 102
Superior concha, of ethmoid bone, 20
Superior constrictor muscle, 117, 119, 120, 143
  function of, 117
  innervation of, 143
  structure of, 117
Superior fossa, of arytenoid cartilage, 93, 96, 113
Superior horn, of thyroid cartilage, 90, 93, 97
Superior longitudinal muscle, 135, 144
  function of, 135
  innervation on, 144
  structure of, 135
Superior mental spine, of mandible, 34, 134
Superior nuchal line, of occipital bone, 15, 53, 72
Suprahyoid muscles, 99, 100, 102
Supramastoid line, of temporal bone, 17
  see also Inferior temporal line of temporal bone
Surface
  anterolateral, of arytenoid cartilage, 91, 93
  dorsomedial, of arytenoid cartilage, 90, 91, 98
  ethmoidal, of palatine bone, 20, 29, 30
  facial, of maxillary bone, 25, 25
  infratemporal of maxillary bone, 25
  medial, of arytenoid cartilage, 91
  nasal
    of lacrimal bone, 28
    of maxillary bone, 25, 27, 29
  orbital
    of lacrimal bone, 28
    of maxillary bone, 20, 25, 27, 32
    of sphenoid bone, 21, 24
Swallowing, 116, 117, 119, 127

Symphysis
   of mandible, 34, 99, 100, 141
   of pubis, 43
System, organ, 3

T

Tail, rudimentary, 36
Teeth
   canine, 27
   central incisor, 27
   incisor, 34
   lateral incisor, 27
   mandibular, 34
   maxillary, 27
   molar, 34
   second premolar, 34
Temporal bone, 15, 17, 18, 22, 24, 31, 35, 53, 99, 102, 120, 122, 131, 133
Temporal line, inferior
   of frontal bone, 7, 131
   of parietal bone, 15, 131
   of temporal bone, 17, 131
Temporal process, of zygomatic bone, 18, 30
Temporalis muscle, 131, 144
   function of, 131
   innervation of, 144
   structure of, 131
Temporomandibular joint, 35
Tendinous intersections, of rectus abdominis muscle, 64
Tendon
   central
      of diaphragm, 69
      of omohyoid muscle, 102
   intermediate, of digastric muscle, 99, 102
   tendinous intersections, of rectus abdominis muscle, 64
Tensor palatini muscle, 122, 143
   function of, 122
   innervation of, 143
   structure of, 122
Thoracic cavity, 69
Thoracic vertebra, 36, 40, 41, 49, 73, 77, 79, 82
Thorax, 56, 51, 59, 62, 82
   dimensions of, 51, 71
   muscles of, 56
Thyroarytenoid muscle, 110, 111, 113, 115
   function of, 111, 113
   innervation of, 115
   structure of, 111
Thyroepiglottic muscle, 113, 114, 115
   function of, 114
   innervation of, 115
   structure of, 114
Thyrohyoid ligament, 97
Thyrohyoid membrane, 93, 97

Thyrohyoid muscle, 104, 115
   function of, 104
   innervation on, 115
   structure of, 104
Thyroid cartilage, 88, 89, 90, 94, 96, 97, 104, 106, 109, 111, 113, 114, 120, 126
Thyroid notch, of thyroid cartilage, 90
Tissue, 3
   type of, 3, 4
      bone, 3
      cartilage, 3
      connective, 3, 116, 134
      epithelial, 4, 82
      muscle, 3, 4, 116
      nervous, 4
Tongue, 90, 116, 117, 127, 133, 134, 135, 136, 137, 138, 143, 144
   innervation of, 143, 144
Torus tuberius, of auditory tube, 128
Trachea, 82, 87
Tracheal cartilage, 82, 87
Tracheal rings, 82, 87
Transverse, 4
Transverse abdominal muscle, 69, 83, 84
   function of, 69
   innervation of, 84
   structure of, 69
Transverse arytenoid muscle, 107, 108, 111, 115
   function on, 108
   innervation of, 115
   structure of, 107
Transverse foramen, of typical cervical vertebra, 39
Transverse muscle, 137, 138, 144
   function of, 138
   innervation of, 144
   structure of, 137
Transverse process
   of eleventh and twelfth thoracic vertebra, 41
   of first cervical vertebra, 39
   of first thoracic vertebra, 40
   of lumbar vertebra, 81
   of typical cervical vertebra, 39, 56, 77, 79
   of typical thoracic vertebra, 40, 49, 51, 79
   of typical vertebra, 38
Transversus thoracis muscle, 62, 63
   see also Triangularis sterni muscle
Trapezius muscle, 72, 73, 83, 85
   function of, 73
   innervation of, 85
   structure of, 72
Triangularis muscle, 141, 142, 144
   function of, 142
   innervation of, 144
   structure of, 141, 142

Triangularis sterni muscle, 62, 63, 84
  function of, 63
  innervation of, 84
  structure of, 62, 63
Trunk, 53, 61, 66, 77, 82, 69
Tubercle
  anterior, of first cervical vertebra, 39, 116
  posterior, of first cervical vertebra, 39
  of rib, 40, 51, 61, 79
Tympanic portion, of temporal bone, 17

U

Uncinate process, of ethmoid bone, 20, 33
Uvula, 121, 122
  see also Azygos uvulus muscle

V

Velopharyngeal closure, 116
Velum, 116, 117, 120, 121, 122, 126, 127
Vena cava, 69
Ventral, 4
Ventricle, of larynx, 86, 110
Ventricular fold, of larynx, 87, 96, 113
Ventricular ligament, 96
Ventricular muscle, 113, 115
  function of, 113
  innervation on, 115
  structure of, 113
Vertebra
  cervical, 16, 36, 39, 40, 54, 53, 56, 73, 77, 79, 82
  lumbar, 42, 36, 71, 73, 77, 81
  sacral, 36, 73
  thoracic, 36, 40, 41, 42, 49, 73, 77, 79, 82
  typical, 36, 38, 49
Vertebral border, of scapula, 48, 60
Vertebral column, 7, 36, 48, 49, 51, 53, 54, 65, 77, 79, 80, 81, 82

Vertebral foramen, of typical vertebra, 36
Vertical crest, of nasal bone, 32, 32
Vertical muscle, 136, 137, 144
  function of, 137
  innervation of, 144
  structure of, 136
Vertical part, of palatine bone, 28, 29
Vestibule, of larynx, 86, 96
Vocal fold, of larynx, 86, 100, 109, 113
Vocal ligament, 96, 109
Vocal process, of arytenoid cartilage, 91, 94, 109
Vocalis muscle, 109, 110, 113 115
  function of, 110
  innervation of, 115
  structure of, 109
Voice production, 86
Voluntary muscle, 3, 4
Vomer bone, 20, 24, 27, 30, 31

X

Xiphoid process, of sternum, 47, 69

Z

Zygomatic arch, 25, 31, 128
Zygomatic bone, 13, 18, 24, 25, 27, 31, 32, 128, 139
Zygomatic head, of quadratus labii superior muscle, 139, 141
Zygomatic process
  of frontal bone, 7, 13, 31
  of maxillary bone, 25, 27, 32
  of temporal bone, 17, 18, 32
Zygomaticus muscle, 139, 141, 144
  function of, 141
  innervation of, 144
  structure of, 139, 141